Popular Mechanics

Trim Carpentry

Rick Peters

HEARST BOOKS
A division of Sterling Publishing Co., Inc.

New York / London
www.sterlingpublishing.com

Photography: Christopher J. Vendetta
Page Layout: Sandy Freeman
Illustrations: Bob Crimi
Contributing Editor: Cheryl Romano
Copy Editor: Barbara McIntosh Webb
Indexer: Nan Badgett

Safety Note: Homes built prior to 1978 may have been constructed with hazardous materials: lead and asbestos. You can test painted surfaces with a test kit available at most hardware stores. Asbestos can be found in ceiling and wall materials, joint compound, insulation, and flooring. Hire a professional, licensed hazardous-removal company to check for this, and remove any hazardous materials found.

Library of Congress Cataloging-in-Publication Data

Peters, Rick
  Popular mechanics. Trim carpentry / Rick Peters.
    p. cm.
  Includes bibliographical references and index.
  ISBN-13: 978-1-58816-687-6 (alk. paper)
  1. Trim carpentry. I. Title. II. Title: Trim carpentry.
  TH5695.P48 2008
  694'.6—dc22
                                      2007–042465
10 9 8 7 6 5 4 3 2 1

Published by Hearst Books
A Division of Sterling Publishing Co., Inc.
387 Park Avenue South, New York, NY 10016

Popular Mechanics and Hearst Books are trademarks of Hearst Communications, Inc.

www.popularmechanics.com

For information about custom editions, special sales, premium and corporate purchases, please contact Sterling Special Sales Department at 800-805-5489 or specialsales@sterlingpublishing.com.

Distributed in Canada by Sterling Publishing
c/o Canadian Manda Group, 165 Dufferin Street,
Toronto, Ontario, Canada M6K 3H6

Distributed in Australia by Capricorn Link (Australia) Pty. Ltd.
P.O. Box 704, Windsor, NSW 2756, Australia

Manufactured in China

Sterling ISBN 978-1-58816-687-6

CONTENTS

Acknowledgments 4

Introduction 5

Trim Basics 6

Trim Carpentry Toolbox 26

Trim Carpentry Know-How 46

Floor and Wall Trim 90

Window and Door Trim 128

Ceiling Trim 142

Architectural Trim 164

Index 190

Metric Equivalency Chart 192

Photo Credits 192

ACKNOWLEDGEMENTS

For all their help, advice, and support, I offer special thanks to:

Justin Robinson with Decorative Concepts for supplying photos, technical information, and their superbly crafted fireplace mantel kits.

Kathy Ziprick with Fypon for supplying lightweight and easy-to-install urethane foam trim used throughout this book.

Elina Gorlenkova with Armstrong Ceilings for providing their Easy Up grid system and ceiling paneling.

Jason Feldner of Bosch Power Tools and Accessories for technical assistance and for supplying the superbly engineered miter saws and accessories used throughout this book.

Haley Burch with EasyCoper Tool Company, Inc., for supplying their ingenious EasyCoper accessory for the saber saw that lets you cope trim like a pro.

Karen Slatter at Benchdog for supplying technical information and images, as well as their easy-to-use crown molding jig.

Brad Witt of Woodhaven for providing technical assistance and images, as well as their handy crown molding jig.

Fred Gunzner of Avenger Products for their company's add-on laser guide systems that let anyone cut faster and more accurately with any miter saw.

Christopher Vendetta, ace photographer, for taking great photos in less than desirable conditions.

Bob Crimi for his superb illustrations.

Sandy Freeman for her excellent page layout talents that are evident in every page of this book.

Barb Webb, copyediting whiz, for ferreting out mistakes and gently suggesting corrections.

Heartfelt thanks to my constant inspiration: Cheryl, Lynne, Will, and Beth.

4

INTRODUCTION

If your home was built in the last 20 years or so, its basic flavor is all too likely to be plain vanilla. As builders struggle to cut costs, one of the first things to go is the very thing that makes a home unique—the trimwork, inside and out. When the same window and door casings and baseboards are used throughout, uniformity rules.

Fortunately, you can change this by replacing the old trim or by installing trim where there was none before—such as adding a surround to a fireplace. The difference can be staggering: In a weekend, a room can go from boring to beautiful. All it takes is some trim, a little patience, and some know-how.

In this book, we'll take you through everything you'll need to be able to install trim like a pro. Chapters 1, 2, and 3 cover trim basics, trim carpentry tools, and trim carpentry know-how. The chapters that follow are all about trimming out specific spaces: floor and wall trim, window and door trim, ceiling trim, and architectural trim. Each trim project features clear, step-by-step photography, along with detailed instructions and illustrations.

With today's new trim products and materials, you don't have to settle for plain vanilla. Pistachio, anyone?

James Meigs
Editor-in-Chief, **Popular Mechanics**

5

CHAPTER 1:
TRIM BASICS

"Cookie-cutter" homes tend to look that way because of trim—there isn't much of it, and what there is looks like every other home on the block. Like personal accessories, interior and exterior trim gives a room or a house its distinctive look. One of the best ways to add style and grace to any home, inside or out, is to add trim.

But trim does more than just look good. It also conceals and seals gaps around doors, windows, floors, and ceilings. In this chapter we define types of trim and various materials, accessories, and finishing supplies. With this knowledge you'll be able to select the right trim for your home to make it uniquely yours.

THE FUNCTION OF TRIM

Trim may be decorative, practical, or both. In most cases, practical trim is also decorative.

DECORATIVE TRIM

Decorative trim serves no practical use, outside of adding visual interest to a room or exterior. Although the ceiling brackets in the photo below appear to support the ceiling, they're actually made out of urethane foam and offer no support at all. Their sole purpose—which they do rather well—is to add a distinctive touch to an otherwise plain ceiling. Likewise, the medallions affixed to the corners of the arched doorway are also purely decorative. Other examples of decorative trim include wall frames (pages 100–101), ceiling medallions (like the one shown in the top right photo on page 10), and trim used to dress up cabinets (pages 116–119). ▼

PRACTICAL TRIM

Trim that's practical is usually designed to either conceal gaps or seal gaps, as described below. Other practical trim includes wainscoting, like that shown in the photo at right. Not only does this trim protect the walls, but it also serves as a back for the built-in seating. Another practical way to protect walls is with chair rail (pages 98–99). Practical trim can also serve other functions, such as displaying china on a built-in plate rail or adding storage space with a built-in niche, as described on pages 182–183.

CONCEALS GAPS

The most common trim in home interiors is designed to conceal transition gaps in materials between floors, walls, ceilings, and window and door openings. These include baseboard to conceal the gap between the wall and floor covering, and window and door casing to conceal the gaps between the wall coverings and the window and door jambs. ▼

PRACTICAL TRIM

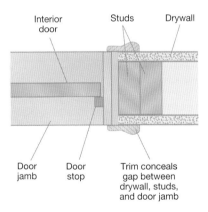

Interior door   Studs   Drywall

Door jamb   Door stop   Trim conceals gap between drywall, studs, and door jamb

SEALS OPENINGS IN WALLS

When trim is attached to an exterior door or window, it also serves to help create a weather-tight seal. For starters, it holds in or conceals the insulation that's installed in the gap between the window or door jamb and the framing. Once in place, the edges of the trim are sealed with silicone caulk to further help prevent drafts and keep out moisture. ▼

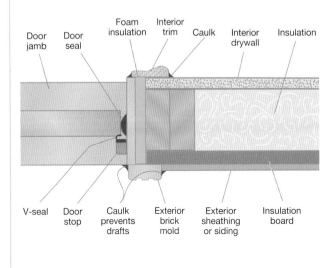

Door jamb   Door seal   Foam insulation   Interior trim   Caulk   Interior drywall   Insulation

V-seal   Door stop   Caulk prevents drafts   Exterior brick mold   Exterior sheathing or siding   Insulation board

Ceiling trim

One of the most often overlooked areas in a home, regarding trim, is the ceiling. That's too bad, because ceiling trim can add tremendous impact to a room, as shown in the photos on this page. Crown molding, ceiling medallions, and period trim are commonly used for ceiling trim.

CROWN MOLDING

Crown molding is the most popular of all ceiling moldings. It serves as a transition from wall to ceiling and can add substantial visual interest to the room. (See pages 150–157 for directions on how to install crown molding.) ▼

PERIOD TRIM

Transitional trim between ceilings and walls is selected to match a style period (such as Baroque or Early American). It's purely decorative in nature but adds much to the room. ▼

CEILING MEDALLIONS

Although the primary role of a ceiling medallion is decorative, it can also conceal damage around an electrical box. Ceiling medallions range in sizes from around 12" in diameter up to a whopping 6 feet. (See page 148 for how to install a ceiling medallion.)

Floor trim

Baseboard is the primary trim used to conceal the gaps between floor and wall coverings. You'll find baseboard in three basic flavors: one-piece, built-up, and period.

ONE-PIECE BASEBOARD

One-piece baseboard consists of a single piece of trim and comes in a wide variety of profiles. It is the easiest type to install (for step-by-step directions on installing one-piece baseboard, see pages 92–95). ▼

BUILT-UP BASEBOARD

PERIOD BASEBOARD

Baseboard that's manufactured to mimic a specific period style can add a lot to a room. Note how well the tall, two-piece baseboard blends in with the Arts & Crafts–style room below. ▼

BUILT-UP BASEBOARD

Baseboard that consists of more than one piece is commonly called a built-up baseboard. Although more complex to install than a single piece, it offers a much wider selection of profiles as you can mix and match trim to create the look you're after. (See pages 96–97 for more on installing built-up baseboard.)

Window and door trim

There are two basic types of interior window and door trim: picture frame casing, and stool-and-apron, as described below. Exterior trim can vary widely, ranging from simple casing to elaborate façades (see pages 170–172). Similarly, door trim varies from simple to complex (see pages 173–175 for an example of an elaborate door façade).

PICTURE FRAME CASING

When a window is framed with casing that's mitered at its corners, it's called picture frame. This is the simplest but also plainest type of window casing. ▼

DOOR TRIM

Door trim is usually installed like picture frame window trim except that there's no bottom piece. So the sides are cut square at their bottom ends to rest on the flooring, as shown. ▼

STOOL-AND-APRON CASING

Adding a horizontal stool-and-apron to the bottom of a window adds visual interest. For more on stool-and-apron trim, see page 132.

Wall trim

Wall trim has many functions. Chair rail and wainscoting protect walls. Wall trim can also be used to dress up an ordinary fireplace or plain walls.

WAINSCOTING

Wainscoting protects walls, add visual interest, and does an excellent job of dividing up wall space. The two most common types of wainscoting are tongue-and-groove (pages 102–109), and frame-and-panel, as shown below and described on pages 110–115. ▼

FIREPLACE SURROUNDS

Technically, a fireplace mantel is trim installed above the fireplace and is commonly used to display knick-knacks or family photos. When many folks think of a fireplace mantel, they're really thinking of a fireplace surround, and envision trim that wraps fully around the fireplace. This trim can be simple or elaborate. (See pages 184–189 for directions on how to install a fireplace surround.)

WALL FRAMES

Wall frames are basically mitered picture frames of trim attached to walls to break up large expanses of walls or add visual interest. (See pages 100–101 for directions on installing wall frames.) ▼

FIREPLACE SURROUND

TYPES OF TRIM

Trim is often classified by its function: baseboard, wainscoting, chair rail, window-and-door casing, ceiling trim, decorative, and practical. Multiple-piece trim is further defined as either built-up or interchangeable, as described on page 17.

BASEBOARD

Baseboard conceals the gaps between wall and floor coverings. It is secured to both the sill plate and the wall studs. ▼

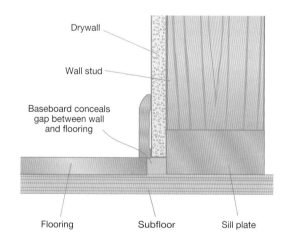

Drywall

Wall stud

Baseboard conceals gap between wall and flooring

Flooring

Subfloor

Sill plate

WAINSCOTING

Wainscoting is a partial wall covering that can begin at any height and terminates at the flooring. Wainscoting is attached to either backer strips or panels (see page 104) and is either tongue-and-groove or frame-and-panel. ▼

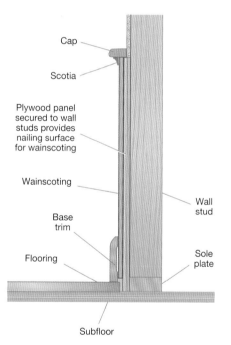

Cap

Scotia

Plywood panel secured to wall studs provides nailing surface for wainscoting

Wainscoting

Wall stud

Base trim

Flooring

Sole plate

Subfloor

CHAIR RAIL

Chair rail is typically a single piece of trim that runs horizontally around a room. It's located roughly one-third of the way up the wall. Although its purpose is to protect the wall from chairs, it is also a great way to separate sections of wall. Many homeowners decorate the wall above and below chair rail differently, sometimes painting one and wallpapering the other. ▼

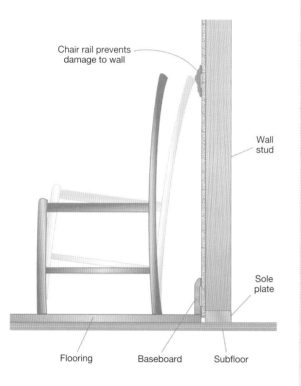

WINDOW-AND-DOOR CASING

Window-and-door casing conceals the gaps between framing and jambs and also provides a convenient way to create a seal against the elements, as illustrated in the top right drawing. Casing usually attaches to the framing and to the window or door jambs.

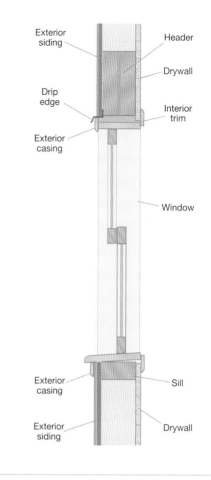

CEILING TRIM

The most common ceiling molding is crown molding. It's commonly secured to either a foundation strip or a backer block. Other ceiling trim includes non-crown trim, coffered or paneled ceilings, and decorative brackets. ▼

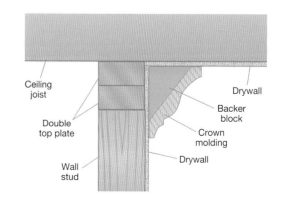

15

DECORATIVE TRIM

Common trim that's purely decorative includes medallions, rosettes, wall frames, imitation pilasters, and columns and brackets. Many of these trim pieces are made of urethane foam and offer no structural support. ▼

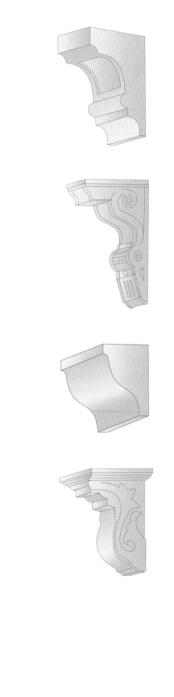

PRACTICAL TRIM

Two common types of practical trim are window and door stops and screen stops. Most window and door stops are just rectangular strips of wood that are attached to the jambs around the perimeter of a window or door. They are there simply to keep the door or window in place. Screen stop is trim that's applied over the screening attached to a wood frame. Yes, it can make the screen look better, but its main purpose is to hold the screening in place. ▼

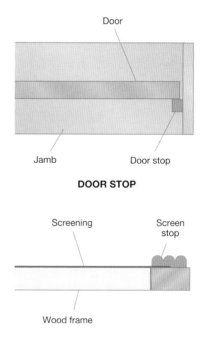

16

BUILT-UP TRIM

Almost any trim can be made up of multiple pieces either glued or nailed together. The advantage that built-up trim offers over a single piece is that the possibilities are virtually endless by mixing and matching different profiles. Built-up trim is often the only way to duplicate an existing molding, as described on pages 86–87. This technique also allows you to make custom and one-of-a-kind trim. ▼

INTERCHANGEABLE TRIM

Some trim manufacturers offer special two-piece trim that accepts different accent strips. This lets them make a single molding that can offer different looks. The two most common of these are crown molding and under-cabinet molding. This helps keep the manufacturers' costs down so they can pass the savings on to you. You'll most often find this type of trim offered by cabinet manufacturers. ▼

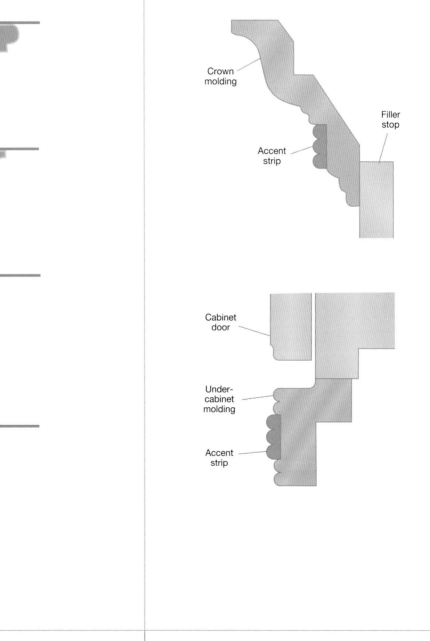

Crown molding

Filler stop

Accent strip

Cabinet door

Under-cabinet molding

Accent strip

17

TRIM AND MOLDING MATERIALS

At your local home center or lumberyard, you'll find trim made from softwood, hardwood, medium-density fiberboard, foam, and plastic. Softwood trim is broken down into two grades: stain-grade and paint-grade, as described below.

STAIN-GRADE TRIM

Trim that can be stained or finished with a clear top-coat is classified as stain-grade and is made from a single, solid piece of wood. Because it's made from a single piece, stain-grade trim is much more expensive than paint-grade. Buy it only when you plan on applying a stain or a clear topcoat. ▼

PAINT-GRADE TRIM

Most wood trim on the market is paint-grade. Instead of being milled from a single piece of wood, many smaller pieces are joined together with a finger joint. Using smaller pieces like this does a couple of things. It makes the trim cheaper to man-ufacture, and it also helps reduce warp and twist caused by variations in grain. On the downside, if you were to apply a stain or topcoat to a piece of paint-grade trim, the variations in wood and also the joints would be quite evident. ▼

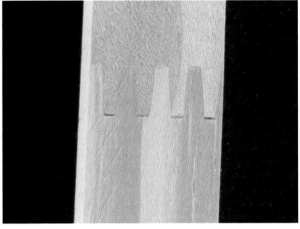

MDF

Because it's expensive, both hardwood and softwood trim is gradually being replaced with alternative mate-rials. One of these materials is MDF (medium-density fiberboard). Just like the sheet stock (described on page 21), trim made of MDF is stable, takes paint well, and won't warp or twist. It's usually sold pre-primed and ready for paint. ▼

FOAM

Another popular alternative to wood trim is foam. When foam trim first hit the market years ago, it wasn't well received. That's because the density of the foam was low. But recent advances in materials and production technology allow for better foam trim. In fact, today's foam offers many advantages over wood. It machines easily, it's flexible, it's stable (no warp or twist whatsoever), and it's basically impervious to weather. Some trim manufacturers, like Fypon (www.fypon.com), even offer stainable foam molding that's textured like wood. Once stained, you'd swear it's natural wood.

LAMINATED TRIM

Some foam trim manufacturers offer trim covered with a variety of laminates, including plastic, paper, and foil coverings. Wood-grained coverings are the most popular, but unlike real wood, these cannot be stained—so make sure you choose the exact color that you're after.

PLASTIC

The most weatherproof trim is plastic. Plastic molding is made from PVC (polyvinyl chloride) like its plumbing cousin, PVC pipe. It is impervious to water and so is great for exteriors or damp locations. Its only drawback is color: It comes in white only, and it's tough to finish in other colors.

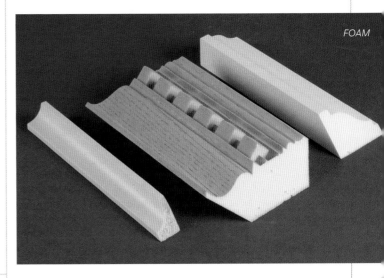

FOAM

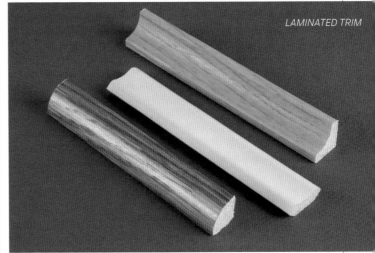

LAMINATED TRIM

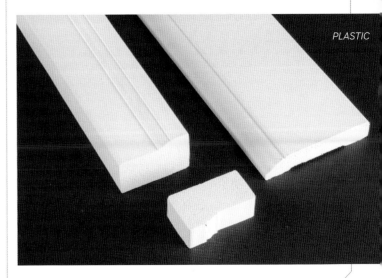

PLASTIC

HARDWOOD AND SOFTWOOD LUMBER

Many trim jobs call for lengths of hardwood or softwood lumber. Examples of this include foundation base for crown molding (page 73), and the frame portions of frame-and-panel wainscoting (as described on pages 110–115). To shop wisely, you should know about the different grades of hardwood and softwood lumber.

HARDWOOD

The grades of hardwood you'll encounter are FAS (firsts and seconds), select, No. 1 common, and No. 2 common. Basically, the grade of a piece of hardwood depends on how much clear wood the board will yield in relation to its total square footage or surface measure. Each grade specifies this as a percentage: roughly 83% for FAS and select, around 67% for No. 1 common, and 50% for No. 2 common. One important thing to realize about grading is that the clear wood in a No. 2 common board is the same quality as that in an FAS board—there's just less of it. ▼

SOFTWOOD

Most softwood lumber is used in the construction industry. That's why softwood grading takes into account strength, stiffness, and other mechanical properties. The problem is that no two woods have identical characteristics. This means that every softwood species has its own set of grading guidelines. For trim jobs where the wood will be painted, go with D select; for clear topcoats, choose either B&BTR or Superior grades. ▼

SHEET GOODS

Occasionally a trim job—particularly when installing wainscoting—will require the use of sheet goods. Common sheet goods are plywood, MDF (medium-density fiberboard), and hardboard.

PLYWOOD

As with lumber, it's good to know about the different grades of softwood and hardwood plywood available. Grade in softwood plywood refers to the quality of the veneer used for the face and back veneers (A-B, B-C, etc.). Stick with AC plywood and place the A face out. The different grades of hardwood plywood describe only its appearance, not its core type or strength. Choose hardwood plywood only if you're using a clear topcoat, then pick a species to match your trim—oak and birch plywood are commonly available. ▼

MDF

Medium-density fiberboard (MDF) is an engineered product where wood fibers are coated with resin and then heat-compressed to form sheets. Since there's no grain, changes in humidity have little effect on MDF. And this means stability. The uniformly small particles also create a solid, homogenous edge that takes paint well and machines easily. ▼

HARDBOARD

Hard, dense, and relatively flat, hardboard is commonly called Masonite, the brand name of the leading manufacturer. Hardboard is made in much the same way as MDF—finely ground processed wood and resins are bonded together under heat and pressure. It comes in 4 x 8-foot sheets in $1/8$" and $1/4$" thicknesses. ▼

TRIM ACCESSORIES

To make it easier to install trim, most home centers and lumberyards offer trim accessories like plinth blocks, rosettes, and other pre-made corners.

PLINTH BLOCKS

A plinth block is a rectangular block that serves as the base for a vertical trim piece such as door casing or a pilaster. (See pages 140–141 for more on using and installing plinth blocks.) ▼

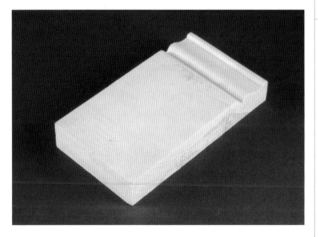

ROSETTES

A rosette is a square block of wood used to handle vertical to horizontal trim transitions, such as when trimming around a door opening. A rosette usually has a design carved or stamped in its center.

ROSETTES

PRE-MADE CORNERS

If you need to wrap trim around a corner but are leery about cutting miters to join the pieces, consider using pre-made corners. Both inside and outside corners are commonly available in a variety of shapes, sizes, and materials. ▼

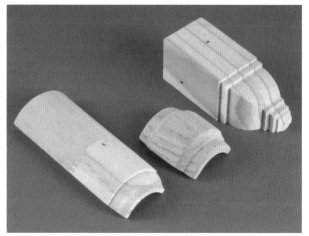

SEALANTS

When trim is installed around windows and doors, it's usually sealed to help prevent drafts. The most common materials used for this are caulk, caulk saver rod, and expanding foam. Caulk can also conceal gaps between trim and adjacent walls.

CAULK

The types of caulk you'll use to seal trim are acrylic latex and silicone. Acrylic latex can be painted and is used to fill gaps between trim and interior walls. Silicone caulk seals gaps on exterior walls. Make sure to always use 100% silicone caulk for maximum flexibility. ▼

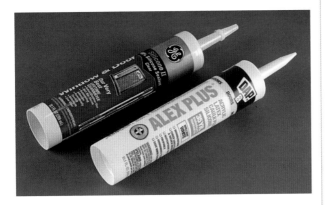

CAULK SAVER ROD

If you encounter a large gap that needs to be filled with caulk, don't pump a lot of caulk into the gap. Instead, consider first pressing some foam rod (often called backer rod or caulk saver) into the gap before applying caulk.

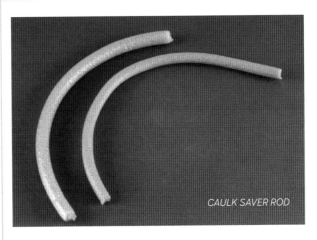

CAULK SAVER ROD

EXPANDING FOAM

Today, expanding foam is the number one way to fill gaps around windows and doors. It's important that you choose the correct type, however. Standard expanding foam will expand to roughly twice its original size. It's best used to fill gaps around framing members, and it dries rigid. Window-and-door foam is specially formulated to expand less than standard foam—and it's more flexible when dry. While standard foam tends to bow window and door jambs, window-and-door foam does not. ▶

ADHESIVES

Most trim is held in place solely with nails. But sometimes you'll find that this isn't possible. So it's a good idea to further secure the trim with an adhesive. The three most widely used adhesives for this are carpenter's glue, construction adhesive, and polyurethane glue.

CARPENTER'S GLUE

Carpenter's or yellow glue is a cross-linked PVA (polyvinyl acetate). Carpenter's glue sets up quickly (about 15 minutes), and clamps need to be applied only for about 1 hour.

CONSTRUCTION ADHESIVE

Construction adhesive fills gaps, sets up without clamps, and is surprisingly strong. It typically comes in tubes and is applied with a caulk gun. Urethane foam is best secured with a high-quality polyurethane construction adhesive.

POLYURETHANE GLUE

Polyurethane glue is a relative newcomer to the adhesive market. It's extremely versatile, and it works particularly well with foam and plastic trim. It does have its disadvantages, though. If it gets on your skin, there's not a solvent around that'll take it off—make sure to wear rubber gloves. And, although a bead of polyurethane glue will expand 2 to 3 times its size to fill gaps, the dried foamy substance has no strength.

CARPENTER'S GLUE

CONSTRUCTION ADHESIVE

POLYURETHANE GLUE

FINISHING SUPPLIES

Your choices for finishing trim include paint, stain, and clear topcoats. You'll also need to deal with filling nail holes in trim.

PAINT

Paint is your best bet when you want to cover up trim (such as paint grade trim), or you want to add color to your trim. Always start with an oil- or alkyd-based primer before applying the top coat of paint.

PAINT

STAIN AND CLEAR TOPCOATS

Solid-wood trim can be left natural or stained. You can either tint wood with semi-transparent stains to allow the wood's grain be seen, or use an opaque stain for a complete cover-up. Whether stained or left natural, the trim should be sealed with a clear topcoat. Clear coats include polyurethanes, spar varnish, and marine varnish. Each of these penetrates into the wood and lets the natural beauty show through.

STAIN AND CLEAR TOPCOATS

PUTTY AND WAX CRAYONS

Putty and wax crayons are used to fill nail holes. There are two types of putty: hardening and non-hardening. Hardening putty is applied prior to finishing and is usually sanded smooth. Non-hardening is color-matched to the finish on your trim (usually natural or stained wood) and is used after the finish has been applied—it does not harden over time. In lieu of putty, some trim carpenters and most cabinet installers use wax crayons to fill nail holes. This is done after the finish is applied, as most cabinet trim is pre-finished. Wax crayons are available in a variety of colors and shades and are easy to use (see page 82).

PUTTY AND WAX CRAYONS

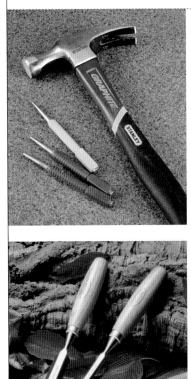

CHAPTER 2:

TRIM CARPENTRY TOOLBOX

Look into any trim carpenter's toolbox and you'll see a mix of hand and power tools. A pro knows that although power tools can save you time, it's the hand tools that often are the most critical. In this chapter we'll make recommendations for both types. For power tools, we'll cover drills, saws, and air-powered tools. Hand tools include measuring and marking tools, as well as tools for fine-tuning trim joints, like planes, files, and chisels. In short, it's everything you'll need to outfit your toolbox for trim work.

LAYOUT AND MEASURING TOOLS

The foundation of a successful trim job is accurate layout and measurement. You can have the fanciest laser-guided chop saw, but if you don't accurately measure and lay out your trim, it won't fit. The tools you'll find useful include: a tape measure, folding rule, level, angle square, angle gauge, digital protractor, framing and try squares, compass, stud finder, chalk line, and laser level.

TAPE MEASURE

A tape measure is the measuring tool for trim work. All tape measures feature a spring-loaded tape that's concave to keep it rigid when extended. Tape measures come in in 8', 12', 16', and 25' lengths. We recommend having a 12' and a larger 25' tape measure on hand to handle both small and large jobs. ▼

FOLDING RULE

Before the advent of the tape measure, the folding rule was used for layout and measuring (middle photo). Folding rules are usually made of wood with brass-protected ends. The old standby was the four-fold, 2' rule that fit easily in a pocket. A more modern version is the folding zigzag rule. We often reach for a folding rule when measuring inside dimensions, or anytime the flexibility of a tape would cause a measurement error. ▶

LEVELS

Trim should be installed level and plumb; the tool for checking this is the level. Levels come in 3', 4', 6', and 8' lengths, and their bodies can be made of wood, metal, or plastic. Most levels have multiple vials, which are curved glass or plastic tubes filled with alcohol (hence the name "spirit" level). A bubble of air trapped in the vial always floats to the highest point on the curve. Smaller "torpedo" levels are handy for checking in smaller spaces where full-sized levels can't fit. ▼

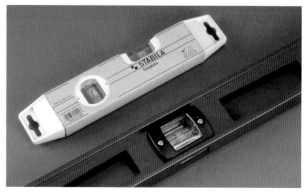

ANGLE SQUARE

An angle square is basically a right triangle made of metal (typically aluminum). A lip on one edge makes it easy to align the square to one edge of the workpiece when marking square or miter cuts. Angle squares are also handy for guiding cuts with a circular saw (see page 62). ▼

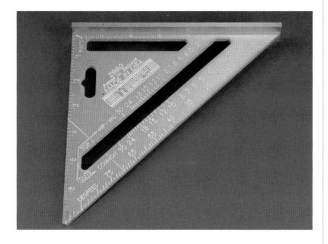

ANGLE GAUGE

An angle gauge is a specialized tool designed for measuring the angle of two adjacent surfaces. Plastic angle gauges like the one shown below take all the guesswork out of cutting trim to fit against out-of-square surfaces. ▼

PRO-TIP: DIGITAL PROTRACTOR

Many of the walls, floors, and ceilings in a home are not square, plumb, or level. The challenge to installing trim is to identify the actual angles where walls, ceilings, and floors meet. Some help: Bosch recently introduced the Miterfinder digital protractor. This tool is actually four tools in one: an angle finder, a compound-cut calculator, a protractor, and a level. (For more on using this handy tool, see page 60.) ▼

FRAMING SQUARE

The long legs of a framing square make it ideal for checking larger surfaces, where a smaller try square or combination square won't give an accurate reading. Framing squares can be found in a number of materials and sizes. We prefer an aluminum square over a steel one, as it's much lighter and won't rust over time. ▼

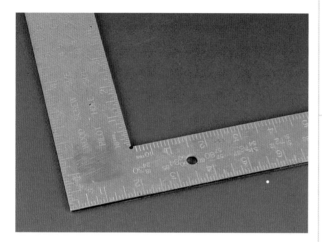

COMBINATION AND TRY SQUARES

A try square is useful for laying out trim for cuts. It consists of a metal blade and either a wood or metal stock (right tool in top right photo). A combination square (left tool in top right photo) is a metal rule with a groove in it that accepts a pin in the head of the square. The head has two faces—one at 90 degrees and the other at 45 degrees. When the knurled nut on the end of the pin is tightened, the head locks the rule in place at the desired location. Combination squares are great for laying out offsets and can also be used as a try square or a depth gauge (see page 51).

COMPASS

Although a compass is useful for laying out circles and arcs, the reason you'll find it in a trim carpenter's toolbox is that it's used to scribe parts to fit, as described on pages 56–57. The most common type is the wing compass, as shown below. The legs of a wing compass are hinged at the top or joined with a spring and forced open or closed by turning a knurled knob that attaches to a threaded post spanning the legs. One leg sports a steel point; the other, a pencil. ▼

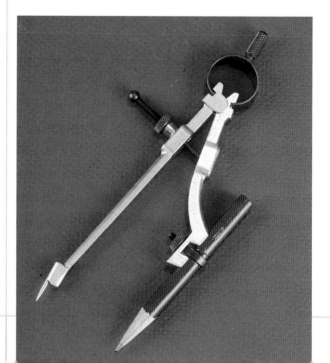

ELECTRONIC STUD FINDER

Before attaching any trim to a wall or ceiling, you'll first need to locate the underlying framing so you can nail the trim to it. An electronic stud finder is the best tool for this job. Recent advancements in technology have driven down the price of these tools while boosting their accuracy. ▼

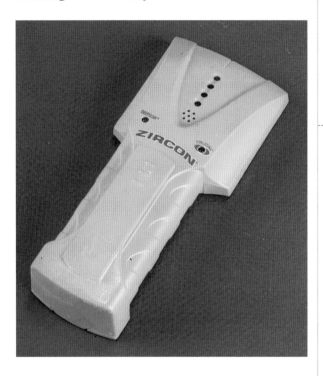

CHALK LINE

PRO-TIP: **LASER LEVELS**

Some trim jobs, like installing chair rail or wainscoting, require drawing a line around the perimeter of a room for positioning the trim. You could snap a chalk line, but then you'd have to remove the chalk from your walls. A better method is to use a laser level, like the one shown below. These are coming down in cost, and can be rented at most home and rental centers. The big advantage of a laser level is that it shoots a perfectly level line along a wall—or around the perimeter of a room—without any marks. ▼

CHALK LINE

You'll find that using a chalk line is the easiest and most accurate way to mark long, straight lines. The case of a chalk line has an opening for chalk. In use, you shake the case to distribute chalk on the line and then pull out the line. A metal hook on the end makes it easy to hook one end at your mark. Then stretch the string to the opposite end and pull the line straight up a couple of inches; let go to "snap" a line. A handle on the side lets you reel the string back into the case, much like a fishing reel.

HAND-POWERED CUTTING TOOLS

Before power tools, all trim was installed by hand. While it still can be, usually trim work is best tackled with a combination of hand and power tools. The hand tools you'll find useful include: a back saw, a coping saw, a miter box, planes, chisels, files, and rasps.

BACK SAW

A back saw is ideal for cutting trim to length. The rigid, thin blade cuts a fine kerf and makes it easy to follow a pencil line. Back saws are identifiable by the brass or steel "back" that wraps around the top of the blade. Back saws come in a variety of lengths, ranging from 8" to 14", and have fine teeth (14 to 22 teeth per inch) that may be sharpened as crosscut, rip, or combination teeth. ▼

COPING SAW

Coping saws are a special type of hand saw where the blade is held in a tensioned frame that is adjustable. Coping saws get their name from what they were designed for—cutting a coped joint, as described on pages 70–71. Besides their thin blades, coping saws excel at making coping cuts because the blade can be pivoted to make it easier to follow a curve. Most coping saws accept a $6^5/8$" very thin, narrow blade with fine teeth—typically 15 to 32 teeth per inch. ▼

MITER BOX

For the most part, back saws have been replaced in the home workshop by the power miter saw (page 37). To cut accurate angles, back saws are frequently placed in a miter box made of wood or plastic to hold the saw at the correct angle. If the saw is sharp and the miter box is accurate, this type of setup works well for cutting trim. ▼

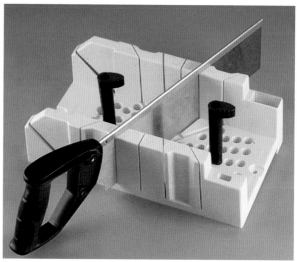

BLOCK PLANE

A block plane is handy for fitting joints and trimming parts to fit. Block planes usually run between 5" and 6½" in length and were originally designed to plane ornery end grain. ▼

HAND PLANE

Hand planes are useful in the shop for jointing edges and all-around smoothing. The two most common types are jack planes and jointer planes, which are very similar in appearance. The big difference is in the length of the sole and the width of the blade. Generally, the cutoff between a jack and a jointer plane is 18". Jointer planes begin around 22"; a mid-sized plane, often called a fore plane, resides in the no-man's land between the two. A common rule of thumb is the longer the sole, the wider the blade. ▶

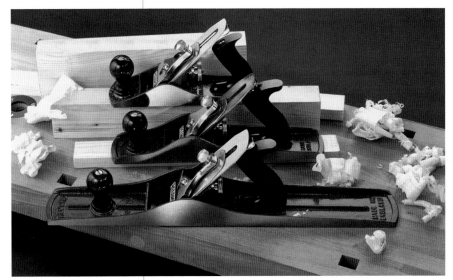

PRO-TIP: SHOOTING BOARD

Even in the hands of a pro, planing a perfectly square edge on the end of a piece of trim is a challenge. It's easy for the plane to tip or tilt, resulting in an uneven cut. That's where a shooting board can come to the rescue. A shooting board is a shop-made jig that lets you hold the work square so that you can accurately trim it with a plane. A shooting board can be made to trim 90-degree cuts or 45-degree miters, as shown in the bottom photo. For more on using a shooting board, see page 81. ▼

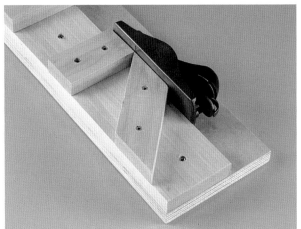

CHISELS

Chisels are useful for fine-tuning trim to fit. They're especially handy for paring away excess wood of a coped joint. Most chisels are similar in appearance. They each have a wood or plastic handle and a blade with a tang or socket. What sets them apart is primarily the profile of the blade and their respective grinds or cutting angles. ▼

WOOD RASPS AND FILES

As with chisels, files and rasps excel at cleaning up and fine-tuning trim to fit. Files come in a huge variety of shapes and sizes. The most common shapes are: mill (or flat), half-round, round, 4-in-hand, and triangular. For all-around work, the half-round shape is your best bet. It combines a gently curving face with a flat face; this combination will handle most jobs. Although rasps closely resemble files, they're very different. Instead of single rows of teeth cut into the metal surface at an angle, rasps have tiny individual teeth in parallel rows. And unlike the smooth cut of a file, a rasp is designed to virtually tear out chucks of wood. This makes them very aggressive, and a rasp in the hands of a seasoned user can remove a lot of material in no time flat.

WOOD RASPS AND FILES

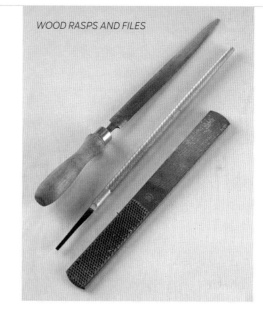

PRO-TIP: RIFLERS

Riflers are specialty files used primarily by carvers to smooth out small details in their work. But they're also excellent for fine-tuning trim. They may be double-ended or come with a handle. Riflers are available individually or in sets, and can be either files or rasps. ▼

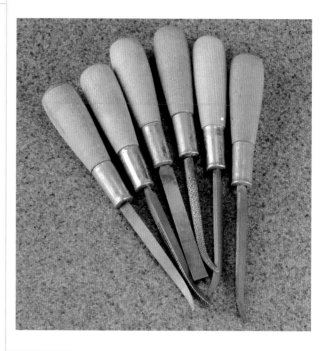

POWER TOOLS

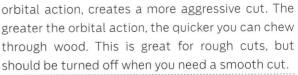

Blades powered by motors require no elbow grease, and features like laser guides allow for precision joints. Power tools for trim include: a circular saw, saber saw, table saw, router, reciprocating saw, biscuit jointer, and chop saw.

CIRCULAR SAW

By far the most common type of circular saw is the shaft-drive. A shaft-drive saw uses the motor's shaft to drive the blade directly. For the average homeowner, this type of saw is all you'll need. Pros find that worm-drive saws permit more accurate cuts. That's because the motor is in line with the arm—it's a more natural motion, and the saw tends to almost guide itself in a straight line in use. But beware: These saws

are both expensive and quite heavy. ▼

SABER SAW

Saber saws are used in trim work to cope joints and trim scribed parts to fit. There are two basic types: standard and orbital. Although inexpensive, a standard saber saw offers limited features. It typically will have only one speed and no orbital action. On an orbital saber saw, the blade can move both up and down and back and forth. As the orbital switch is increased, the blade begins to pivot out during the cut. This pivoting, or orbital action, creates a more aggressive cut. The greater the orbital action, the quicker you can chew through wood. This is great for rough cuts, but should be turned off when you need a smooth cut.

PRO-TIP: EASYCOPER

Many homeowners shy away from coping molding because it takes a steady hand, patience, and considerable skill. But a new accessory called the EasyCoper (www.easycoper.com) brings accurate coping within everyone's reach. EasyCoper is a plastic aid designed by a professional carpenter that simplifies the coping of crown molding. The jig lets you use a saber saw instead of a coping saw to cope the molding. It holds both the molding and the saw at the perfect angle to make the cut, as shown in the bottom photo.

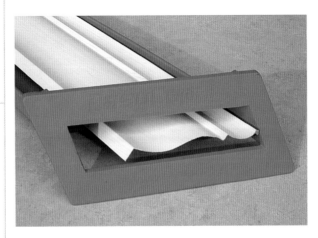

TABLE SAW

A table saw is used for ripping, crosscutting, miters, bevels, joinery—even shaping edges. You'll find one useful for cutting sheet stock and trim to width and length. There are three main types to choose from: bench-tops, contractor's saws, and cabinet saws. A bench-top saw like the one shown below is a good choice for a homeowner with limited space. These saws typically have smaller motors and are direct-drive, making them quite portable and perfect for most trim jobs. ▼

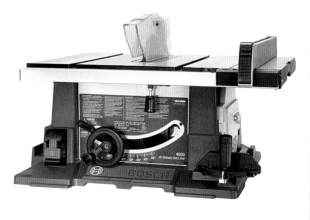

PORTABLE ROUTER

If a trim job calls for custom parts, you'll find a portable router the best tool for shaping edges to create matching profiles. A router also lets you create your own trim profiles, as described on pages 86–87. Routers fall into two basic categories: standard and plunge routers. A standard router consists of a router base and a motor unit. The motor unit slides up and down or rotates within the base to adjust the depth of cut. Plunge routers (top right photo) are similar to standard routers except that the motor unit slides up and down on a pair of spring-loaded metal rods. The normal resting place of the motor unit is at the top of the rods. To make a cut, you release a lever and push down to lower (or plunge) the bit into the workpiece.

PRO-TIP: ROUTER TABLES

Mounting a router in a table effectively turns it into a mini-shaper. You can accurately cut joints, work safely with small parts, and add precision to your router work. There's quite a variety of router tables on the market, everything from bench-top tables and stationary models to shop-made versions. The table shown below is manufactured by Bench Dog Tools (www.benchdog.com). ▼

RECIPROCATING SAW

One of the best all-around demolition tools available is the reciprocating saw (commonly referred to by the brand name Sawzall), as shown in the top photo. It makes quick work of cutting through wall coverings and framing. When fitted with a demolition blade, it's the perfect tool for freeing windows and doors from rough openings prior to their removal. ▼

BISCUIT JOINTER

Looking to join trim parts together quickly and accurately? Consider a biscuit jointer (often called a plate jointer), as shown in the bottom left photo. These specialized tools cut half-moon-shaped slots in wood to accept compressed-wood biscuits. Biscuit joinery is quick and easy and is a great way to strengthen an otherwise weak butt joint, such as when making the frame for frame-and-panel wainscoting, as described on pages 110–115. ▼

POWER MITER SAW

The power miter saw is the trim carpenter's number one tool. It can cut through trim like butter and make precise miter and compound miter cuts with ease. There are three main types of power miter saw: chop saws, compound saws, and sliding compound saws. A chop saw gets its name from its action—it chops boards into smaller pieces across their width. Most of these saws are capable of making simple miter cuts between 0 and 45 degrees—usually in both directions. A compound miter saw has the ability to tilt the blade. Couple this with the ability to rotate the blade, and you can cut compound miters. A sliding compound saw (bottom right photo) offers all the features of the compound saw, but has a saw carriage that slides, allowing you to cut wider boards. Most have crosscut capacities of at least 12", they're extremely accurate, and they're portable. ▼

PRO-TIP: MITER SAW ACCESSORIES

There are a number of accessories available for your miter saw that can add to its precision and functionality. These include: laser guides, stands, and crown molding jigs.

Laser guides. A laser guide shoots a red laser beam directly onto the workpiece to accurately show where the blade will cut. This accessory has become so popular that many new saws come with laser guides installed as standard. Not to worry, though, if you have an older saw—a number of saw and accessory manufacturers sell laser upgrade kits. Laser guides can shoot a single beam or a dual beam, and are either turned on and off via a switch or are automatically turned on when the blade reaches a set rpm. The guides shown here are made by Avenger Products (www.avengerproducts.com). ▼

SAW STANDS

Saw stands. A number of tool manufacturers make portable stands for miter saws, like the Bosch stand shown in the top right photo. These range from simple fold-up stands to elaborate units with built-in fences and stops, like the one shown here.

Crown molding jigs. There are numerous crown molding jigs on the market that take the guesswork out of cutting crown. The jig shown below is made by Bench Dog Tools (www.benchdog.com); it holds the molding at its intended angle while you cut it. So, no more compound cuts and tedious trial and error. Bench Dog's Crown-Cut jig requires only a simple 45-degree miter cut that any miter saw can handle. ▼

ASSEMBLY TOOLS

The tools you need to attach trim are few and simple: a hammer, nail sets, a driver/drill, and a putty knife to fill nail holes.

Hammer and nail set

Even if you primarily use air nailers (pages 40–42) to attach trim, you'll still need a good hammer and some nail sets to handle the occasional spot where an air nailer won't fit, and for demolition work such as prying apart parts. Hammer weights range from a delicate 7 ounces up to a beefy 28 ounces—the most common size is 16 ounces and works well for trim work. One of the secrets to nailing effectively is knowing when to stop driving the nail with a hammer. The closer you get to the surface, the greater the chance of dinging it with the hammer. To eliminate the chance of dings, stop when the head of the nail nears the surface. Then reach for a nail set, as shown below. A nail set is designed to finish the job by driving the nail below the surface, or "setting" it. ▼

DRIVER/DRILL

Although most trim is secured with nails, many trim foundations—like backer panels and backer blocks—are attached with screws. Other trim work, like wrapping a post with a column (page 169), requires screws. And the driver/drill is the best tool for the job. You'll also use one to drill pilot holes in the ends of trim to prevent the nails from splitting the trim as they're driven in. ▼

PUTTY KNIFE

Putty knives are indispensable for trim work. They'll protect walls when you're removing trim (page 89), and are the best way to apply putty to fill nail holes. You'll find them with metal or plastic blades. Plastic blade knives are best used on pre-finished trim, where a metal blade could scratch the finish. ▼

AIR-POWERED TOOLS

Next to the power miter saw, the air-powered nailer was the biggest boon to trim carpenters. An air-powered nailer can drive and set a nail, brad, or staple in the blink of an eye. And compared to hammers, air-powered nailers have less of a tendency to damage the surface of the trim. Most air-powered tools require a compressor, a hose, and fittings, as described below (some do not; see page 42). Common air tools used for trim work include: brad nailers, narrow-crown staplers, and finish nailers.

COMPRESSORS

Most air tools are powered by a compressor. There are two basic types: oil-lubricated and self-lubricating. Most trim work can easily be handled with a small self-lubricating compressor. A portable compressor (around 1½-hp) with a 4- to 6-gallon tank will do the job. Oil-free or "self-lubricating" compressors run without oil. Instead, they use non-metal piston rings, Teflon-coated parts, and sealed bearings. ▼

PRO-TIP: COMMON COMPRESSOR TERMS

Air tank: a metal tank attached to a compressor to store air. Common types are: pancake, cylindrical, and twin. Capacities range from 4 gallons to over 60 gallons.

CFM (cubic feet per minute): the volume of air being delivered by a compressor to an air tool, and a measure of the compressor capabilities; it's the actual amount of air in cubic feet that a compressor can pump in 1 minute at working pressure.

Compressed air: free air that has been pressed into a volume smaller than it normally occupies. As compressed air exerts pressure, it performs work when released and allowed to expand to its normal free state.

Oil-lubricated compressor: a type of compressor that uses oil to lubricate the internal parts; consists of an air pump powered by either a gas engine or an electric motor, and a tank to store the compressed air.

Oil-free compressor: a type of compressor that uses non-metal piston rings and Teflon-coated parts instead of oil to keep things running smoothly.

Portable compressor: any compressor that features a compressor and motor so mounted that they may be easily moved as a unit.

PSI (pounds per square inch): the measure of air pressure or force delivered to an air tool by the compressor. Most air tools require 90 PSI.

AIR HOSES

There are two basic types of air hose to connect an air tool to a compressor: standard and lightweight. Relatively inexpensive, standard PVC hose resists both oils and sunlight, and remains flexible over a wide range of temperatures. Lengths vary from 25 up to 100 feet. Popular in construction, plastic hose is rugged and lightweight. If you've ever lugged a heavy hose around all day, you'll appreciate the lighter weight. Note that this hose isn't as flexible as standard hose and has an annoying habit of not lying flat—it often gets tangled around your feet.

FITTINGS

The most common fitting used for air hoses and tools is a brass threaded fitting. Common sizes are $1/4$" and $3/8$" NPT threads. Quick-connect fittings let you connect and disconnect tools from the air line without having to shut down the compressor. That's because the female half of the male-female coupling pair has a built-in shutoff valve. This makes the female coupling considerably more expensive than the male. Although quick-connect fittings are typically sold in pairs of male and female couplings, they're also available individually. ▼

BRAD NAILER

A brad nailer is the air tool of choice for attaching thin or fragile trim. They typically shoot 18-gauge brads that can vary in length from $5/8$" to $1\frac{1}{2}$". Just pick the right length for the job and load it in the magazine. ▼

NARROW-CROWN STAPLER

Narrow-crown staplers shoot $1/4$" or $3/8$" staples, in gauges ranging from 16 to 22. Staples offer greater holding power than brads and nails. The major drawback to using a staple is its footprint. Since it's difficult to hide the head of a staple, they're primarily used in areas that won't be seen. ▼

FINISH NAILER

A finish nailer drives 15- and 16-gauge nails varying in length from $3/4$" to $2^3/4$". This makes them ideal for attaching trim, chair railing, crown molding, window and door casings, etc. When looking to buy a finish nailer, the first choice you need to make is what gauge nail to shoot. Thinner 16-gauge nails are less likely to split wood than the heavier 15-gauge. But the disadvantage to the smaller gauge is that the nails tend to follow the grain in wood and often deflect off course, sometimes protruding out the face. ▼

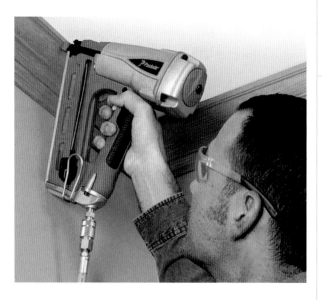

COMPRESSOR-LESS NAILERS

Wouldn't it be nice to have the power and precision of an air nailer but without the compressor and hose? That's what the folks at Paslode (www.paslode.com) have been offering for years (they pioneered cordless air nailers in 1986). A cordless nailer uses a recharge-able battery and a replaceable fuel cell to power a combustion motor to drive fasteners. Cordless nailers are a joy to use. Once you've used one, you'll dread going back to a standard nailer. ▼

COMPRESSOR/NAILER KITS

Many air tool manufacturers are now offering kits that include a compressor, hose and fittings, and one or multiple air nailers, as shown in the top right photo. These "plug and play" packages make it easy to get into air power, as everything you'll need is included in the kit. Note that some kits come with fasteners, while some do not.

TRIM FASTENERS

Nails are sized using the antiquated "penny" system (abbreviated as "d"). This system is based on how much 100 nails used to cost. Penny is now used to indicate the length of the nail; see the illustration below. Nails can be purchased in 1-, 5- and 50-pound boxes, or in any quantity from a retailer that stores their nails in open bins. ▼

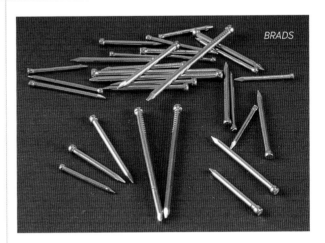
BRADS

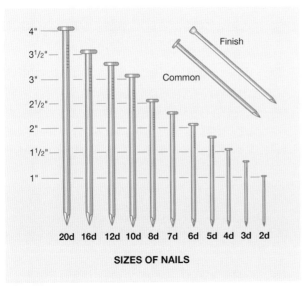

SIZES OF NAILS

FINISH NAILS

Finish nails tend to be thin, and have a cupped head. They are used primarily for attaching trim to framing. Finish nail sizes typically range from 1" in length up to 2". Exterior finish nails are galvanized. ▼

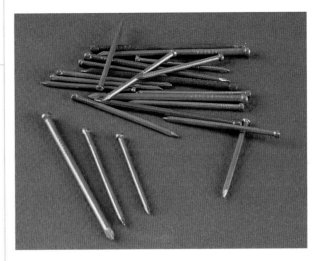

BRADS

Brads, as shown in the top right photo, are most often used to attach delicate trim. Gauge and length vary. You can find brads ranging from 1/2" up to 1 1/2" in length. Gauges are based on wire gauge, and commonly vary from 15- to 18-gauge.

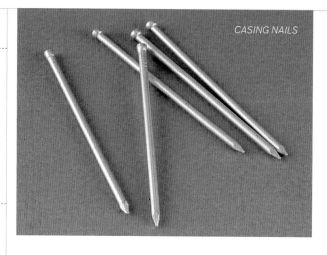

PRO-TIP: WEATHER-RESISTANT FASTENERS

If the nails you'll be using will be exposed to the elements, they should be hot-dipped galvanized, be electroplated galvanized, or be made of a metal that's impervious to moisture, such as aluminum or stainless steel.

CASING NAILS

Casing nails, as shown in the top right photo, look like finish nails except that they are thicker and longer. The head of the nail also tends to be flat versus that of a finish nail. It gets its name from its main use—securing casing to framing.

AIR-POWERED FASTENERS

Pro trim carpenters generally use air nailers to fasten trim in place because they both drive and set the fastener in an instant. Air nailers use special fasteners that come in sticks. These sticks can be straight or angled, as shown below. They also come in standard and exterior finishes. ▼

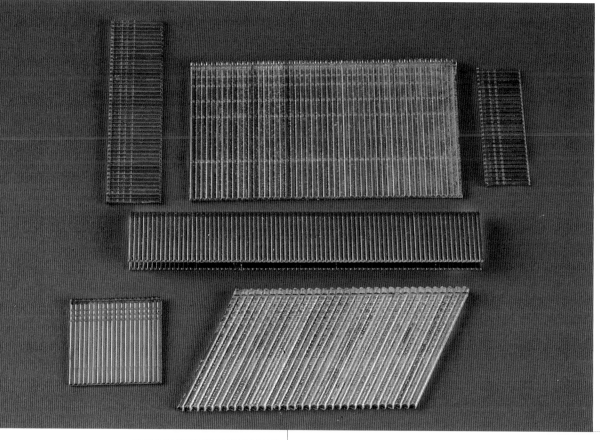

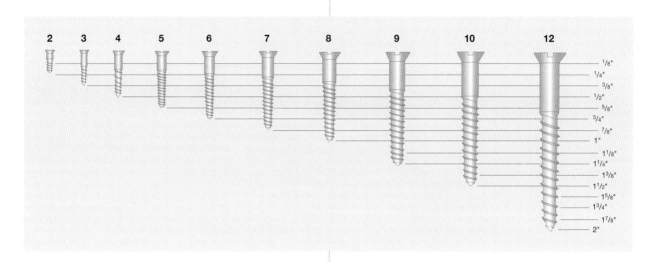

2	
3	1/8"
4	1/4"
5	3/8"
6	1/2"
7	5/8"
8	3/4"
9	7/8"
10	1"
12	1 1/8"

SCREWS

Although nails are used to hold most trim in place, there will be times when you need the extra holding power that a screw offers. Screws can be used to fasten foundation moldings to wall studs, to fasten backer strips to wall studs, and to mount load-bearing trim (like the plate rail described on pages 120–123). Regardless of the screw, whenever possible, select square-drive heads versus slotted or Phillips-head screws. The unique square recess in the screw heads (and matching screwdriver or driver bit) greatly reduces the tendency of a standard bit to "spin" while driving in a screw.

PRO-TIP: TRIM-HEAD SCREWS

There's a special type of screw that's designed for attaching trim—a trim-head screw. On a trim-head screw, the head is roughly half the size of a standard screw head (left screws in the photo below). This small head lets you substitute a screw in place of a nail whenever you need extra holding power.

⚠ WARNING

In the case of left bevel cutting, remove the sub fence. Supposing it is not able to remove it, it Will contact the blade or some part of the tool, causing in serious injury to operator.

CHAPTER 3:

TRIM CARPENTRY KNOW-HOW

So what's to know about trim? You just cut it and nail it in place, right? Technically, this is true. But how long do you cut the trim? At what angle do you cut it to make it fit perfectly? What do you secure the trim to? How do you create invisible joints? All of these questions and more are answered in this chapter. We'll take you though layout and measurement basics, cutting trim, fine-tuning joints, installing trim, and touch-up work—everything you'll need to install trim like a pro.

TAPE MEASURE TIPS

The tape measure is the primary layout and measuring tool for trim jobs. Make sure that you use a quality tape, and use the same tape for your entire job. Using multiple tape measures can inject untraceable errors as you work.

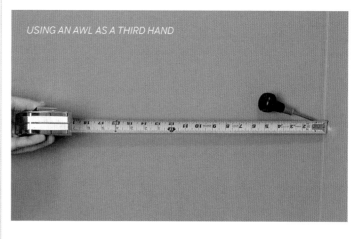

USING AN AWL AS A THIRD HAND

CHECKING A TAPE FOR ACCURACY

When you're buying a tape measure, skip the bargain bin; it pays to buy quality. Go with a name brand that you can trust, like Starrett, Stanley, or Lufkin. And, take the time to check it for accuracy at the store. To do this, simply extend the tape out several feet and bend it back on itself. Align the inch marks and check to see if the graduations align. On cheaper tapes, you'll often find that the graduations don't match up.

USING AN AWL AS A THIRD HAND

Working by yourself and need a "third hand" to hold one end of a tape? Reach for an awl. Just secure the end of a tape measure (or chalk line) to a workpiece with an awl. Then stretch it to the opposite end to measure (or snap a line).

USING THE SLIDING HOOK

Another thing to check for when buying a tape measure is a quality hook that's securely riveted to the end of the tape, while still allowing it to move freely for inside and outside measurements. The reason the hook needs to slide is for inside measurements—it's designed to slide in the exact thickness of the hook to provide an accurate measurement, as illustrated below. ▼

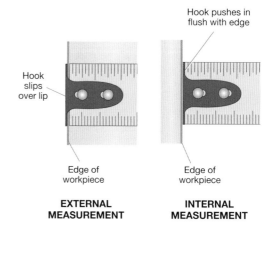

Hook slips over lip

Hook pushes in flush with edge

Edge of workpiece

EXTERNAL MEASUREMENT

Edge of workpiece

INTERNAL MEASUREMENT

USING A FOLDING RULE

Before tape measures, carpenters and cabinet-makers used folding rules. Folding rules are made of wood, with brass pivot joints on their ends. A common older folding rule was the four-fold 2-foot rule that fit easily in a pocket. The more modern type of folding rule is the zigzag rule; it's still widely used because, unlike a tape measure, it will stay rigid when extended. A folding rule can be used like a tape measure to measure the length of an object. Just unfold the rule as long as needed, and read the scale. But where a folding rule really shines is measuring inside corners, as described below.

MEASURING INTO A CORNER

To measure into a corner, begin by unfolding the rule as long as possible. Then butt one end of the rule against a wall or up against an opening (as shown here) and slide the extension out until it contacts the opposite wall. ▼

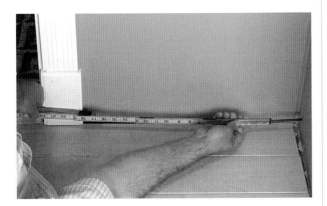

THE EXTENSION

The extension on a folding rule is a metal bar that slides out of one end to make inside measurements a snap. On quality rules, the markings on the extension are stamped into the metal (as shown below); cheaper versions have only printed markings. Once you've butted the extension up against a wall or other object, slide it out carefully and calculate the length (see below). ▼

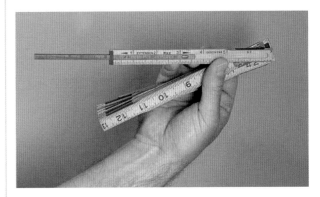

CALCULATING THE LENGTH

To calculate the length of your measurement, simply add the reading of the extension to the extended length of the rule. ▼

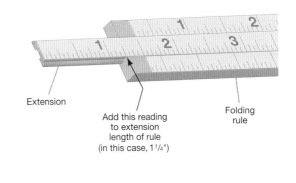

Extension

Add this reading to extension length of rule (in this case, 1¼")

Folding rule

USING A TRY OR COMBINATION SQUARE

Try and combination squares are used in trim jobs to lay out cuts, check parts for square, and mark offsets.

CHECKING FOR ACCURACY

If a try square or combination square isn't accurate, it's not worth much. To check to see whether a square is actually square, place its stock up against a known flat edge. Then draw a line along the blade and flip the stock over. Align the blade with the line you just drew—it should be in perfect alignment. If not, you'll have to adjust either the stock or the blade by filing it down. ▼

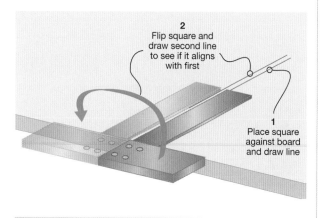

2
Flip square and draw second line to see if it aligns with first

1
Place square against board and draw line

DRAWING ACCURATELY

If you watch a draftsman in action, you'll note that he or she always places the pencil at the exact point where they want to draw a line, then slides a square over so it butts up against the pencil. This might seem obvious, but a lot of people do the opposite—set the square on the mark and then draw the line. The problem with this is that it doesn't take into account the distance between the square and the centerpoint of the pencil lead. ▼

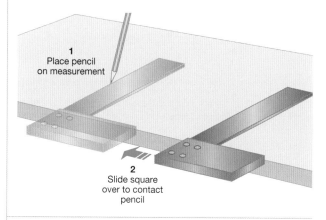

1
Place pencil on measurement

2
Slide square over to contact pencil

CHECKING PARTS FOR SQUARE

Besides layout work, a try or combination square is the perfect tool to check parts for square. The most accurate way to do this is to position the blade of the square on the edge of the workpiece. This lets you position the workpiece so that light can shine through from behind to indicate any gaps. ▼

ADJUSTING A COMBINATION SQUARE

A combination square is a metal rule with a groove in it that accepts a pin in the head of the square. The head has two faces—one at 90 degrees and the other at 45 degrees. When the knurled nut on the end of the pin is tightened, the head locks the rule in place at the desired location. In use, you simply loosen the knurled nut, slide the rule to the desired measurement, and tighten the nut. ▼

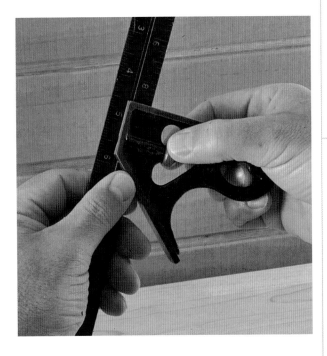

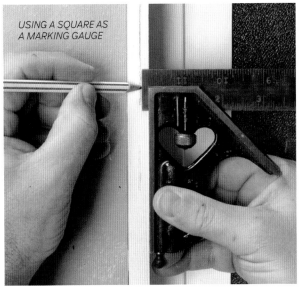

USING A SQUARE AS A MARKING GAUGE

USING A SQUARE AS A DEPTH GAUGE

Another use for a combination square is as a depth gauge. To use a square for this, loosen the knurled knob and place the head of the square on the surface that you'll be measuring from. Then slide the blade over until it bottoms out. Tighten the knurled knob and remove the blade to read the depth. ▼

USING A SQUARE AS A MARKING GAUGE

Say, for instance, that you want to lay out an offset for trim (as shown here) or a series of lines on a fireplace mantel where you're going to rout a set of flutes. Just set the blade the desired distance from the edge of the workpiece, lock it in place, and then butt the head up against the workpiece. You'll note a small notch centered in the end of the metal rule; this is for your pencil. Position the pencil in the notch and, while holding it firmly in place, slide the square along the edge of the workpiece to mark the line, as shown in the top right photo.

USING STORY STICKS

There are two types of story sticks: reusable sticks and one-offs. Both are simply sticks of wood that let you accurately measure and lay out trim without having to read a tape measure. They've been used for centuries, and even with the advent of laser measuring devices, many savvy trim carpenters and cabinet installers still rely on them today.

ple as the metal binder clips shown in the left photo or small C-clamps for thicker stock. In general, the longer the distance you need to measure, the thicker the strips should be so they'll stay rigid.

MEASURING WITH STORY STICKS

To measure using story sticks, first loosen or remove the clamps holding the strips together. Then extend the strips out so each pointed end touches a surface. Now secure or tighten the clamps. The length of the story sticks is the exact distance between the surfaces. ▼

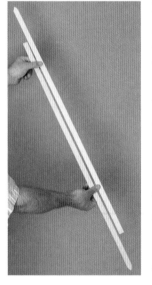

REUSABLE STICKS

Story sticks let you "measure" for trim without a tape. They consist of a pair of sticks that are held together with clamps, as illustrated in the drawing below. The ends are pointed to allow them better access into corners, which results in more accurate measurements. Depending on the thickness of the strips of wood, the clamps can be as sim-

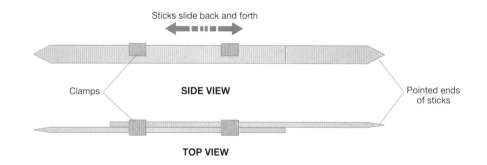

Sticks slide back and forth

Clamps   **SIDE VIEW**   Pointed ends of sticks

TOP VIEW

TRANSFER TO TRIM

Once you've "measured" your space with the story sticks, you can use them to transfer this distance to your trim part. Just place the story sticks next to your trim part with the ends flush, as shown in the top photo, and mark the trim as shown. Cut the trim to length and it'll be a perfect fit...no measurements needed. ▼

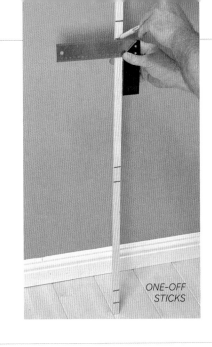

ONE-OFF
STICKS

TRANSFER THE MEASUREMENTS

Now you can use your one-off stick to work your way along a wall to lay out a pattern in seconds. This method is particularly useful when laying out backer strips, as described on page 104, and for locating wall frames, as described on pages 100–101. ▼

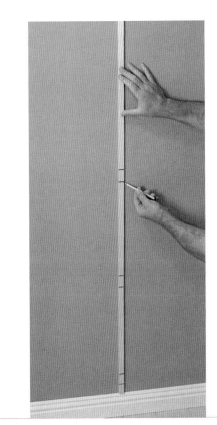

ONE-OFF STICKS

Although accurate measurements are good, one-off story sticks are better for laying out trim, as they remove any chance of misreading a measurement. Basically you're making a reusable template. To make a one-off story stick, lay out your trim pattern on a scrap of wood as shown in the top right photo.

ANGLE SQUARE BASICS

An angle square is a heavy-duty metal layout tool that's basically a right triangle. A lip on one edge makes it easy to mark stock for mitered or square cuts. Angle squares are also handy for guiding cuts with a circular saw (see page 62).

TYPES OF ANGLE SQUARES

There are two types of angle squares: fixed and adjustable. A fixed angle square (foreground in the photo) is a single piece and has one lip at one edge. An adjustable angle square (background in the photo) has a two-piece adjustable bar in place of a lip. The bar is held in place with a pivoting rivet on one end and a knurled knob on the other. The knob passes through an arc milled in the square and threads into a matching bar on the other side. Any angle between 0 and 45 degrees can be set on this square, which makes it much more handy than a fixed angle square, which can only be used for 45 and 90 degrees. ▼

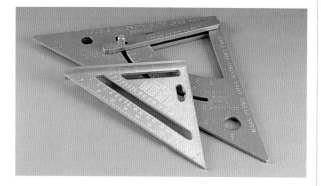

MARKING WITH AN ANGLE SQUARE

Just as you do when using a try or combination square (see page 50), it's best to place your pencil at the exact point where you want to draw a line and then slide the square up so it butts up against the pencil, as shown. If you do the opposite—set the square on the mark and then draw the line—you don't take into account the distance between the square and the centerpoint of the pencil lead, and you'll end up with an inaccurate line. ▼

MITERS AND ANGLES

You can also used a fixed or adjustable angle square to lay out angle cuts. For a fixed square, your only option is 45 degrees. With an adjustable square, you can lay out any angle between 0 and 45 degrees. ▼

USING A FRAMING SQUARE

Unlike the short legs of a try or combination square, the long legs of a framing square make it ideal for checking and laying out larger surfaces. Framing squares can be found in a number of materials and sizes. We prefer an aluminum square to a steel one, as it's much lighter and won't rust over time.

MORE THAN A SQUARE

Most folks don't realize that a framing square is useful for more than just layout and checking for square. Most squares also contain a lot of useful information stamped or printed on their legs. Typical information includes data on laying out rafters (as shown below), wood screw and pilot hole size, and standard lumber and board feet calculations. ▼

ACCURACY MEASUREMENTS

For added accuracy when using a framing square, butt a scrap of stock up against the edge you're checking and then butt your square against the scrap. This ensures that the edge of the square is flush with the edge of your workpiece. Alternatively, allow the blade of the square to hang down over the edge of the workpiece and press it firmly against the edge. This way the square will be aligned with the edge. ▼

SIMPLE ANGLES

You can even use a framing square to lay out simple angles like 45 degrees. Here's how: Just place the square on your workpiece and align the same measurement on its outside edges with the bottom edge of your workpiece. The resulting angle is exactly 45 degrees. ▶

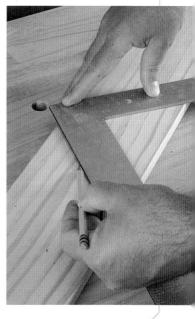

SCRIBING

Depending on how out-of-square a room is, you may need to fine-tune a piece of trim to get it to fit properly. A simple way to mark a piece of trim for this is to "scribe" it. Scribing is a technique that allows you to transfer irregularities of a wall or other part directly onto your trim so you can cut it for a perfect fit. Scribing can be done with a compass or a scrap of cardboard, as described below.

HOW IT WORKS

Scribing uses a pencil spaced a set distance from a reference point to transfer a shape onto your trim, as illustrated below. The reference point can be the pointed end of a compass or the flat edge of a piece of cardboard. ▼

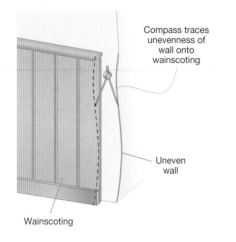

Compass traces unevenness of wall onto wainscoting

Uneven wall

Wainscoting

WITH A COMPASS

To scribe a piece of trim with a compass, place the trim where it will be installed. Then set a compass to slightly wider than the largest gap between the trim and the surface you're scribing. Next, place the compass point on the surface with the pencil tip resting on the trim. Now slide the compass along the surface. The pencil will scribe the unevenness of the surface directly onto the trim. ▶

WITH A CARDBOARD SCRAP

To scribe a part with a scrap of cardboard, cut a piece of cardboard to span the trim and the surface that you're scribing. Then poke a small hole in the cardboard for your pencil at just slightly wider than the largest gap between the trim and the surface that you're scribing. Now you can scribe the part as you would with a compass, except that you'll be running the edge of the cardboard along the surface to be scribed. ▶

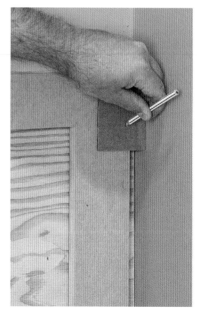

CUT THE SCRIBED PART

Once you've scribed the part, you can cut it to match the scribed profile. Gradual curves and tapers can be cut with a hand saw, as shown here. Tighter curves and more intricate profiles are best cut with a saber saw or a coping saw. Note that if you're scribing the end of a long piece of trim (such as crown molding), it's best to cut the trim long, scribe and cut, and then cut the trim to final length. ▼

CHECK THE FIT

After cutting the scribed part, test the fit by butting the part up against the surface that you scribed, as shown in the top right photo. It should fit perfectly. If it doesn't, fine-tune the fit as described at right.

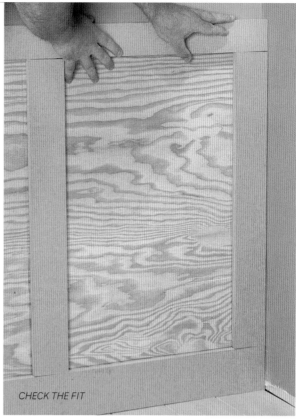

CHECK THE FIT

FINE-TUNE AS NEEDED

What tool or tools you use to fine-tune the fit of a scribed part will depend on the shape of the scribed edge. For gradual curves and tapers, use a block plane (page 33). For tighter curves and intricate profiles, use a file and rasps or riflers, as described on page 34. ▶

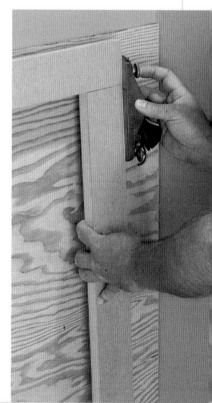

LAYOUT TIPS

Accurate layout and measurement is vital to a professional-looking trim job. Here are some layout tips that you can use to add precision to your work.

USING A RULE STOP

A common cause of inaccuracy when using a metal rule is to measure in from the edge of a workpiece. The problem is positioning the rule so it's absolutely flush with edge. To make sure the rule is flush, just press a scrap of wood firmly against the edge of the workpiece and then butt the end of the rule up against the scrap. ▼

DIVIDING A BOARD EQUALLY

Quick: You need to divide a board into five equal parts. Place one end of a metal rule flush with one edge of the workpiece. Then add 1 to the desired divisible units. Pivot the rule until a number easily divisible by the desired units rests directly over the opposite edge. Then mark to create equal spacing, as illustrated below. ▼

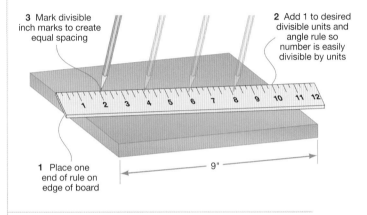

3 Mark divisible inch marks to create equal spacing

2 Add 1 to desired divisible units and angle rule so number is easily divisible by units

1 Place one end of rule on edge of board

9"

FINDING THE CENTER OF A PART

Here's a quick way to find the center of a board with a rule. Position one end of the rule flush with the edge of the workpiece (the scrap block tip mentioned above works great here). Then pivot the rule until a number easily divisible by 2 rests directly over the opposite edge. Divide this number in half and mark the center, as illustrated here. ▼

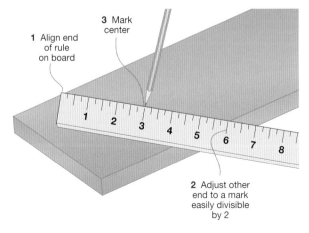

3 Mark center

1 Align end of rule on board

2 Adjust other end to a mark easily divisible by 2

USING AN ANGLE GAUGE

In most homes, what makes mitering trim such a challenge is that many adjoining surfaces are not level, plumb, or square. To miter parts accurately, you need to know the true angle of the intersecting surfaces. There are a couple of ways to do this: with an angle gauge (as described here) and with a digital protractor (as described on page 60).

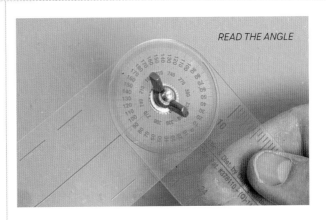

READ THE ANGLE

POSITION THE GAUGE

One of the quickest ways to identify an angle is to use an angle gauge like the one shown in the top right photo. To use an angle gauge, press the blades of the gauge firmly against the surface to be measured and then tighten the tension nut. ▼

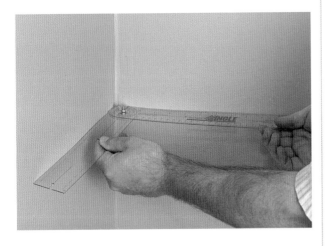

READ THE ANGLE

Pull the gauge carefully away from the wall and read the angle directly off the gauge, as shown in the top right photo.

USING THE ANGLE

There are two ways you can use the angle you just measured. First, you can place the gauge directly on your trim piece, as shown below. Butt one leg up against the long edge of your trim and then mark the cut angle as shown. Alternatively, use the angle to set the miter angle on your power miter saw (as described on page 63) and make the cut. ▼

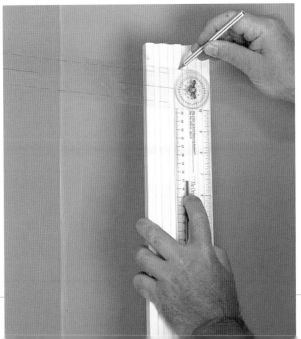

USING A DIGITAL PROTRACTOR

If you're planning on installing a lot of trim, consider purchasing a digital protractor like the Bosch DWM 40L shown here. A digital protractor takes all the guesswork and computations out of installing molding. All you have to do is measure the wall angle with the protractor, and it will compute the miter and bevel angle settings for your saw. Then just set up your saw with the angles provided by the protractor and make your cut to get a perfect fit. What could be easier?

MEASURE THE ANGLE

To use a digital protractor, open up the legs of the protractor and butt each leg flat against the wall surfaces, as shown. Alternatively, you can use the protractor to find the angle on a piece of trim, as shown in the bottom left photo. ▼

STORE THE ANGLE

If your protractor has a "store" or "hold" function, press the hold button to store the wall angle so that you can compute the miter saw settings. ▼

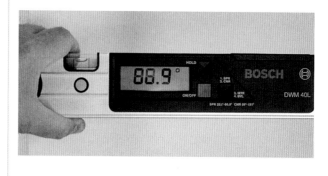

COMPUTE THE ANGLES

For the Bosch protractor shown here, press the BV/MT button and the display will show the value of the miter setting. The miter setting is displayed along with the "MTR" indicator. Make a note of this setting or adjust your saw to this angle. Then press the BV/MT button again to show the bevel angle. The bevel setting will appear along with the "BVL" indicator. Make a note of this setting or adjust your blade to this angle. Your saw is now set up to cut the molding for a perfect fit. ▼

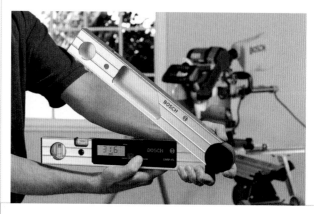

CUTTING TRIM BY HAND

Although most trim installed today is cut with a power miter saw, in days gone by it was always cut to length with a hand saw and a miter box. This technique still works today—it just takes more elbow grease. A miter box holds the trim and positions the saw at the desired cut angle. Manufactured miter boxes have fixed slots at 22½, 45, and 90 degrees. If you make your own miter box, you can cut kerfs in it at any angle.

90-DEGREE CUTS

Because the blade of most back saws is so thin and the kerf is so fine, a well-sharpened back saw will generally steer itself in a perfectly straight cut—all you have to do it start the cut straight. Hold the trim in place (or use built-in clamps, as shown here). Use a light grip on the saw and let it do the work. ▼

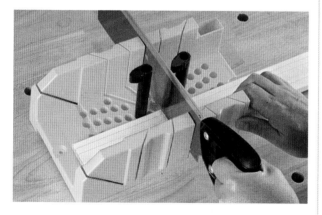

45-DEGREE MITERS

As soon as you angle a blade away from 90 degrees, you're effectively cutting a wider board. That's why it's important to take your time on miter cuts and let the saw do the work. The small, fine teeth of a miter saw or back saw will quickly clog, so stop frequently, lift out the saw, and blow out the sawdust from both the saw kerf and the saw. ▼

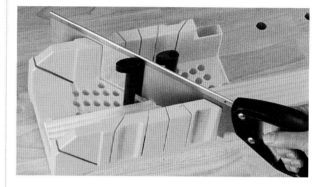

TWEAKING AN ANGLE

All too often you'll find that you need to cut a piece of trim to just slightly off a set angle (like 90 or 45 degrees). On a power miter saw, this is simply a matter of loosening the pivot and angling the blade slightly—not possible on a miter box. A simple solution to this dilemma is to insert a shim between the trim and the box as shown below (playing cards work great for this). Make sure to clamp the workpiece firmly in place before cutting, to prevent it from shifting as you make the cut. ▼

CUTTING TRIM WITH A CIRCULAR SAW

Although you can cut trim to length with a circular saw as described below, you'll get better results with a power miter saw (pages 63–69) or with a hand saw and a miter box, as described on page 61.

CLAMP THE WORKPIECE

Although carpenters frequently cut lumber without using clamps, it's best to always clamp a trim piece in place before making a cut. Carpenters can get away with rough cuts that are slightly out of square, but trim needs to be cut smooth and straight if you want it to look good. ▼

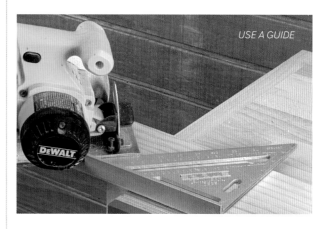

USE A GUIDE

MAKE THE CUT

Then butt the saw up against the adjacent edge of the square and slide the saw and square so the blade is on the waste side of the marked line. Now hold the square firmly in place and depress the trigger. Push the saw through the cut, keeping the saw base in constant contact with the edge of the square. ▼

USE A GUIDE

For added accuracy, use an angle square to guide the saw. Just place it on the workpiece so the lip of the square catches the edge of the workpiece, as shown in the top right photo.

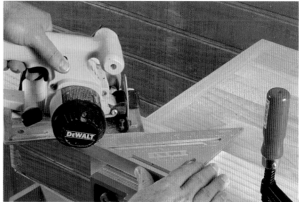

USING A POWER MITER SAW

Power miters saws have made installing trim a lot easier. They've made cutting trim precisely to any angle as simple as setting an angle and lowering the blade, as described below.

SET THE ANGLE

To set the angle on a miter saw, loosen the miter lock—usually the miter handle—and pivot the saw to the desired angle. Then retighten.

ALIGN BLADE WITH MARK

With the miter saw adjusted to the desired angle, position the workpiece for your cut. Start by marking the cut line on your workpiece. Then slide the workpiece over until your mark or line is aligned with the blade. If your saw is equipped with a laser guide, as described on page 65, this is a snap.

CLAMP IN PLACE

One of the best ways to ensure an accurate cut is to clamp your workpiece securely in place before making a cut. Even the best crosscut blade will have a tendency to pull the workpiece slightly as it cuts into the wood. Use either the built-in hold-down clamp or a shop clamp to securely lock the workpiece against the saw table or fence before making the cut.

90-DEGREE CUTS

Crosscutting trim at 90 degrees is simple and straightforward. Position the workpiece and then make the cut. How you position the workpiece will depend on personal preference, and often the size of the workpiece. Depending on size, you may want to place the workpiece flat on the table or hold it vertically against the fence.

SET THE ANGLE

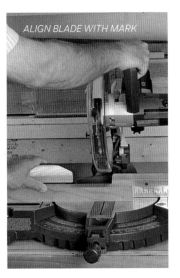

ALIGN BLADE WITH MARK

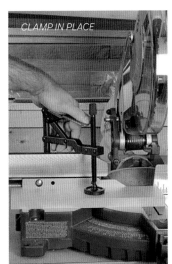

CLAMP IN PLACE

90-DEGREE CUTS

45-DEGREE MITERS

Cutting a workpiece at an angle means that you're technically cutting a wider board. So it just makes sense to slow down the feed rate of the blade as it passes into and through the workpiece. If you cut too fast, odds are that the blade will deflect slightly, resulting in an inaccurate cut. ▼

SAFETY FIRST: MITER SAW SAFETY RULES

1. Secure the miter saw to a stable supporting surface.
2. Use only crosscut miter saw blades that have a zero-degree or negative hook angle.
3. Use only blades of the correct size and type for your saw.
4. Make certain the blade rotates in the correct direction and that the teeth at the bottom of the blade point toward the rear of the saw.
5. Always use the blade guard.
6. Use a sharp, clean blade at all times.
7. Do not use abrasive cutting wheels on your miter saw.
8. Always use the kerf plate or inserts; replace it when damaged.
9. Clean the motor air slots of chips and sawdust to prevent the motor from overheating.
10. Tighten the table clamp handle and any other clamps prior to using the saw.
11. Never start the saw with the blade against the workpiece.
12. Keep arms, hands, and fingers away from the blade to prevent injury.
13. Do not place your hands in the blade area when the saw is connected to power.
14. Don't reach underneath the saw unless it is unplugged.
15. Allow the motor to come to full speed before making the cut. Starting the cut too soon can damage the machine or cause an injury.
16. Never cut ferrous metals or masonry.
17. Do not cut small pieces, as these bring your hand too close to the blade.
18. Never perform freehand cuts. Always hold the work firmly against the fence and table.
19. Properly support long or wide workpieces.
20. Do not allow anyone to ever stand behind the saw.
21. Before operating the saw, check and securely lock the bevel, miter, and sliding fence adjustments.
22. Read the instruction manual before operating your saw.
23. Wear eye and hearing protection. Protect your lungs with a quality dust mask.
24. Maintain your saw in peak condition; keep blades sharp and your saw properly tuned.
25. Keep your work area clean.
26. Keep children and visitors away. Your shop is a potentially dangerous environment.
27. Never reach in back of the saw blade behind the fence with either hand to hold down or support the workpiece, to remove wood scraps, or for any other reason.
28. Never cross your hand over the cut line.
29. Inspect your workpiece before making a cut; if the workpiece is bowed or warped, clamp it with the outside bowed face toward the fence. Always make certain that there is no gap between the workpiece, fence, and table along the line of the cut.
30. Disconnect the power cord before making adjustments or attaching accessories.

PRO-TIP: LASER GUIDES

A laser guide shoots a red laser beam directly onto a trim piece to accurately show where the blade will cut.

Single versus dual beams. A single-beam laser guide shoots a line onto your trim. On most, you can adjust the guide to mark either the right or left edge of the blade. With a dual-beam laser guide, a pair of beams light up both sides of the blade, as shown below—no more guessing where the saw blade will cut. This style of laser guide is handy if you tend to cut on both sides of the blade; that is, your waste cuts can be on either the right or left side of the blade. This isn't so easy with a single-line laser guide, because you need to remember which edge of the blade the laser is marking. It's real easy to end up cutting your workpiece ⅛" too short (the thickness of the blade). ▼

BUILT-IN GUIDES

Built-in guides. Laser guides have become so popular that many new saws come with laser guides installed as standard, like the Hitachi saw shown in the middle photo. Note: Just because the laser in a miter saw laser guide is small doesn't mean it can't be harmful. Laser radiation can hurt you. Make sure never to stare directly into the beam, as it can—and will—damage your eyes.

Add-on guides. If your miter saw isn't equipped with a laser guide and you want to add one, there are a number of accessory manufacturers who make add-on laser guides. These are available with single or dual beams. The guide shown here is made by Avenger Products (www.avengerproducts.com) and is simple to install. All you need to do is replace the outer blade washer with the laser guide unit, as shown below. This type of guide has an internal switch that automatically turns on when the saw blade reaches a set rpm and then turns itself off as the blade stops. ▼

65

POWER MITER CUTS

Regardless of whether you're cutting trim straight or at an angle with a power miter saw, there are three different ways you can position the workpiece: flat on the saw table, vertical against the fence, or tilted against the fence.

WITH THE TRIM AGAINST FENCE

WITH THE TRIM FLAT

The most stable of the positions for cutting trim is laying the trim flat on the table, as this provides the best support. Once you've selected the angle, secure the trim with a stop and/or hold-down and make the cut by bringing the blade down into the trim.

WITH TRIM ANGLED AGAINST FENCE

If you need to make a compound miter cut—such as when cutting crown molding—you'll likely need to angle the trim against the fence for the cut. Some miter saws have built-in stops just for this. If your saw doesn't have built-in stops, you'll need to either clamp a strip across the saw table to hold the trim in place or use a crown molding jig, like the one shown below and described on the opposite page. ▼

WITH THE TRIM AGAINST FENCE

Holding trim vertically on edge is also common, as shown in the top right photo. Trim carpenters find it faster and more convenient to angle the saw to the fence instead of tilting the blade to make a bevel cut. As long as you hold the trim firmly against the fence (securing it with clamps works best), you'll end up with a fairly accurate cut.

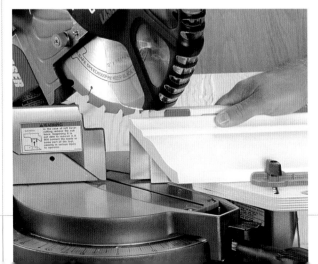

PRO-TIP: CROWN MOLDING JIGS

Crown molding has a well-deserved reputation for being a pain to cut and install. Even if you can handle the mental gymnastics, simply holding the molding itself in place on the saw can be a challenge. Because of this, most saw manufacturers and a number of accessory makers have developed crown molding stops or crown molding jigs.

Built-in stops. Since cutting crown and other trim is such a common task for the miter saw, some manufacturers have added built-in stops to their saws, like the Delta saw shown below. These built-in stops are easy to use—and always right at hand. Just turn the adjustment knob until the stop is roughly positioned. Then lift up the stop until the ears on each side of the stop pop into place. Insert the trim and adjust the stop until the molding is held at the desired angle. ▼

Side stops. Another aid for cutting crown molding is add-on stops like those shown in the top right photo and manufactured by Bosch to fit their saws. The stop is basically a bracket with a lip on the end to hold the crown molding in place. It attaches to the side of the saw base as shown and adjusts in and out to hold a variety of molding widths. This type of stop is simple but effective. ▼

Positioning jig. There are a number of crown molding jigs on the market that are designed to securely hold trim at the desired angle. One such positioning jig shown below is manufactuered by Woodhaven (www.woodhaven.com), and it attaches to their fence system. It also holds trim at the correct angles so all you have to do is make a miter cut. No need to fuss with bevel angles—and it works on either side of the blade. ▼

POWER MITER SAW TIPS

Pros have been using the tips described here for years to add precision to their trim cuts. These include using square and angled stops, shims, kerf inserts, sandpaper to prevent creep, and extensions.

SQUARE STOPS

One of the most reliable ways to prevent creep—the shifting of a workpiece as you make a cut—is to use a stop block like the one shown here. This is particularly effective if you clamp the trim to the table with a hold-down, as shown here. A stop block by itself will keep a piece of trim from being pushed away from the blade—but won't stop the blade from pulling the trim into the blade. For this you need either sandpaper or a hold-down, as described on the opposite page. ▼

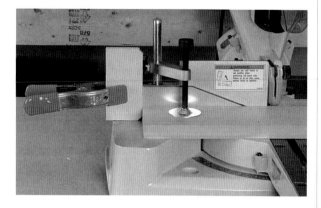

ANGLED STOPS

Square stops work great for 90-degree cuts, but are often found lacking when cutting miters. The ultimate stop for miter cuts is a stop that's angled to match the angled end of the trim, as shown in the top right photo. Make sure to orient the stop as shown so it will capture the angled end against the saw fence. ▼

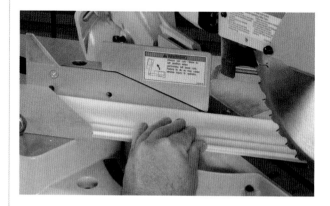

PLAYING CARD SHIMS

Even with careful layout and positioning, you'll often need to tweak the angle of a cut. If you're using a detent and your saw has a lockable detent override, engage it and readjust your miter angle. If your saw doesn't have a detent override, it's virtually impossible to tweak an angle at a detent. One way around this is to shim the trim instead of trying to adjust the saw. A standard playing card works great for this. Just slip it between the trim and the fence (as shown here) to push the end of the trim out slightly and vary the cutting angle by just a hair. If this isn't enough, just add another card. ▼

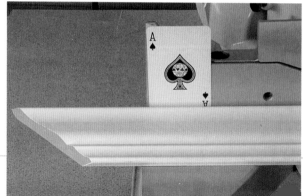

KERF INSERTS

Although chip-out isn't an accuracy problem, it is a real problem when it comes to cutting trim. Chip-out occurs anytime a blade makes an unsupported cut. That is, there's insufficient support under the trim as the blade exits the bottom face of the trim. Without support, wood fibers are torn from the surface. One way to prevent this is with the kerf plate or inserts in the saw top. If your inserts are adjustable, adjust them so they butt up against the blade. This will typically eliminate chip-out. If they're not adjustable, just slip a piece of ¼" hardboard between the trim and the saw table. The hardboard will support the wood fibers of the trim, resulting in a smooth cut. ▼

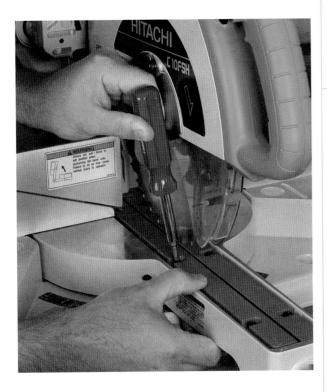

PREVENTING CREEP

One of the most prevalent problems associated when cutting miters—"creep"—can be prevented in a couple of ways. Creep is caused by the blade's tendency to pull or push trim as it makes a cut. To reduce creep, temporarily attach sandpaper to the saw table and/or fence. The grit of the sandpaper "grabs" the trim and keeps it from shifting. Self-adhesive sandpaper works great for this, or you can attach standard sandpaper with rubber cement. ▼

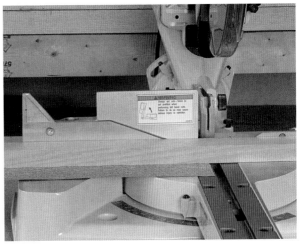

USE AN EXTENSION

Any trim cut will benefit from support—especially long pieces that can tilt and cause an inaccurate cut or personal injury. Depending on the length of your trim, this may be as simple as extending the built-in sliding table extensions of your saw. For longer trim, consider setting up a roller stand or using a miter saw stand with built-in support, as shown on page 38. ▼

COPING MOLDING

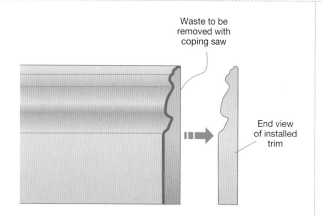

Waste to be removed with coping saw

End view of installed trim

Coped joints are used by pro trim carpenters and cabinet installers who want trim to intersect at inside corners without the gaps normally associated with miter joints. Miter joints are notorious for opening and closing as the humidity changes, producing a varying-width gap throughout the year. Since there's less exposed end grain on a coped joint compared to a miter joint, there's less movement and the joint stays tight. There are two halves to a coped joint, as illustrated in the drawing above. On one half, the molding profile is left intact and simply butted into the corner of a wall or cabinet. The second half is the part that's coped to fit the profile of the molding butted into the corner. When done properly, a coped part will butt cleanly up against the trim profile with no gaps.

EXPOSE THE COPE

The first step to making a coped joint is to install one trim piece so that it butts into the corner. The next step is to expose the cope on the trim to be coped. This can be done easily on the miter saw, as shown in the top right photo. The idea here is to cut the end as if you were doing an inside miter. The miter cut will

expose the wood that needs to be removed to fit against the matching profile of the first piece. ▼

MARK THE COPE

After you've exposed the cope, it's a good idea to highlight and define the profile that you'll be cutting. Just run the tip of a pencil along the profile. This dark line will be a lot easier to follow than just

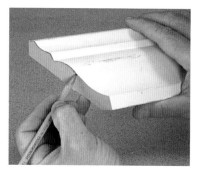

the profile. This is particularly helpful when cutting solid-wood molding, where the face and edge of the exposed miter are the same color.

COPING WITH A COPING SAW

To cope your trim with a coping saw, start by making relief cuts where the profile transitions dramatically, such as a 90-degree turn. Then go back and cut out the waste, taking care not to cut into the face of the molding. What you're after here is a 45-degree cut in the opposite direction of the exposed miter. Beveling the cut in the opposite direction will allow the profile to fit over the molding butted up against the wall. ▼

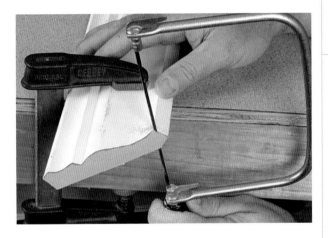

COPING WITH A SCROLL SAW

Normally, waste is removed with a coping saw. But if you own a scroll saw and fit it with a fine blade, it'll make quick work of removing the waste since it's much easier to control the cut, as shown in the top right photo. As with a coping saw, it's best to make a series of relief cuts first where the molding profile changes direction. Then go back and cut out the waste, taking care not to cut into the face of the molding.

COPING WITH A SCROLL SAW

FINE-TUNE THE COPE

After you've removed the exposed waste, try fitting the coped piece against a scrap of the molding. Note any areas where there are gaps, and mark these with a pencil. If there are large gaps, go back to the coping saw and remove the bulk of the waste. If the gaps are small, you can fine-tune the coped profile to fit better. A small round or half-round file works best for this, as shown below. Be sure to angle the file back toward the back of the molding to keep from filing into the exposed profile. Alternatively, a dowel or screwdriver wrapped with sandpaper can quickly remove waste from curved areas. Test the fit frequently, and continue fine-tuning until the coped part fits perfectly against the trim profile. ▼

LOCATING FRAMING MEMBERS

To hold trim in place, fasteners need a solid purchase—either a wall stud, ceiling joist, or top or bottom plate. So the first thing to do on many trim jobs is to locate these. The problem is that they're hidden under a wall covering (typically drywall). There are three ways to locate studs: with a stud finder, by tapping, or by drilling.

STUD FINDER

To use an electronic stud finder, position the finder on the wall, depress the ON button, and gently slide it along, as shown below. The stud finder will have a visual display to indicate that you've found the edges of the stud; some finders also have an audible indicator. ▼

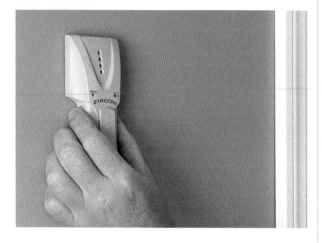

TAPPING

Although crude, tapping works if you've got a good ear. Simply rap a knuckle along the wall slowly and listen for a change in tone, as shown in the top right photo.

Lower tone means no support; higher tone means support and that you're near a stud. Work from both directions toward the higher tone to pinpoint the stud. ▼

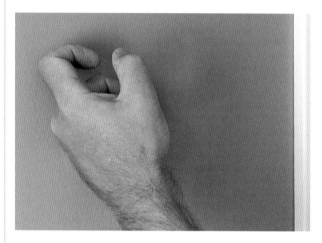

DRILLING

Pro cabinet installers often drill a series of small holes in the wall until they hit a stud. They can get away with this, as they'll be covering the wall with cabinets. If you'll be covering the holes with trim, the same technique can work for you. ▼

TRIM FOUNDATIONS

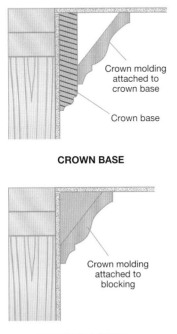

CROWN BASE

Crown molding attached to crown base

Crown base

CROWN BLOCKING

Crown molding attached to blocking

Trim is most often attached to wall studs or ceiling joists. In some cases—especially when installing ceiling trim, like crown molding—it's often best to add extra nailing foundations in the form of base or blocking, as illustrated in the drawing above.

BASE

Base is commonly used for ceiling trim—particularly if the trim is large or heavy, as is often the case with crown molding. The idea is to attach a piece of flat trim to the wall studs (as shown in the top right photo) and then attach the ceiling trim to the base trim, as illustrated in the top drawing. This lets you fasten the ceiling trim to the base anywhere along its length instead of just at the wall studs.

BASE

BLOCKING

Blocking is individual versions of base trim (described at left). Instead of long strips, smaller angled blocks or strips are attached to the studs. This is a popular method for attaching crown, since the angled block provides a larger nailing surface and also supports the trim at the correct angle. ▼

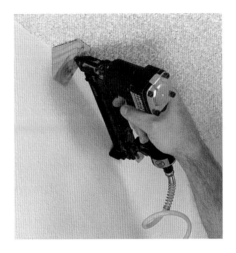

HAMMER BASICS

There are two ways to drive fasteners into trim: with a hammer or with an air nailer. Even if you use an air nailer, you'll still use a hammer for various tasks.

HAMMER GRIP

Surprisingly, even experienced DIY'ers tend to hold a hammer with a weak grip. The most common mistake is to choke up on the handle as if it were a baseball bat. And just as with a baseball bat, this will rob the hammer of any power, greatly reducing its ability to drive a nail. Some might say that this affords better control; but without power, the hammer is useless. It's better to learn to control the hammer with the proper grip. For maximum mechanical advantage from a hammer, grip the handle near the end. Place the end of the handle in the meaty part of your palm, and wrap your fingers around the handle. Avoid a white-knuckle grip, as this will only tire your hand. For less power and a bit more control, position the handle just below the palm, and grip as shown below. This takes the hammer out of alignment with the arm and shoulder, but you may find it more comfortable. ▼

PRO-TIP: HOLDING BRADS

Short brads can be a real challenge to hold in place for nailing without dinging a finger. Here are two simple ways to keep your fingers out of harm's way.

Use a shim. Split the end of a shim, and press a brad into the split to hold it in place. Then position the shim as desired and drive it in. Once it's partially driven in, pull the shim away. ▼

Use needle-nose pliers. If you've got a pair of needle-nose pliers in your toolbox, they make a handy holder for short brads. ▼

CORRECT GRIP

INCORRECT GRIP

USING A NAIL SET

To nail effectively into trim, you need to know when to stop driving the nail with a hammer. The closer you get to the surface, the greater the chance of dinging it with the face of the hammer. To eliminate the chance of dings, stop when the head of the brad or nail nears the surface—usually between $1/16"$ and $1/8"$. Then reach for a nail set to finish the job by driving the nail below the surface, or "setting" it, as shown in the top left photo. Place the set on the nail and tap it gently to set it below the surface. Keep a firm grip on the set, since it's easy for it to slip off and end up being driven into the wood. ▼

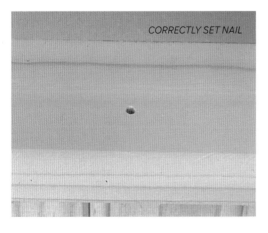

CORRECTLY SET NAIL

PRO-TIP: LOCK-NAILING

Whenever you install mitered trim—even if the joint is perfect—it's a good idea to "lock" the miter joint in place with a nail to prevent it from opening and closing as humidity changes, as illustrated in the drawing below. This technique, called lock-nailing, not only pinches any gaps closed, it also strengthens the joint. When you lock-nail, use a finish nail and make sure to stay away from the very end of the trim, as shown in the bottom right photo, to prevent splitting the trim. ▼

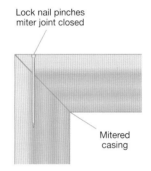

Lock nail pinches miter joint closed

Mitered casing

USING AIR-POWERED TOOLS

For many years, air-powered tools were used only by pros because they were expensive and the average homeowner couldn't justify the expense. But prices have been dropping steadily, and more and more manufacturers are offering economical air nailer kits, as described on page 42. Alternatively, you can purchase a compressor and air nailer separately, or rent them from most any rental center. One of the nice things about trim carpentry is that the nailers (finish and brad) don't require a large compressor, as do framing and coil nailers. Nailers are easy to use, and once you've used one, you'll be hooked.

DISCONNECT AIR BEFORE LOADING

The number one safety rule for using an air nailer is to always wear eye protection. The number two rule is to always disconnect the air hose before loading or removing fasteners. This takes only a second and can prevent a nasty accident. ▼

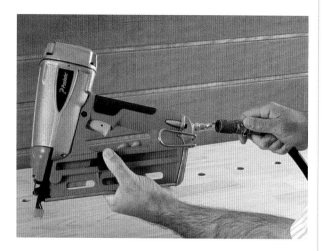

LOADING FASTENERS

There are two common methods for loading fasteners into an air nailer: from the end or from the side, as described below. Air-powered fasteners come in strips and are either straight or angled.

End-loaded fasteners. Pull the magazine follower to the rear of the nailer until it catches. Then slide a strip of nails into the end of the nailer as shown. Push the strip as far forward as possible, and then release the magazine follower. ▶

Side-loaded fasteners. Release the magazine cover and slide it open. Place a strip of nails in the opening, and slide it as far forward as possible. Then slide the cover back in place until it latches. ▶

DEPTH ADJUSTMENTS

With most nailers, you have to fiddle around with the compressor pressure to get the nailer to shoot a nail and set it the correct depth, as shown in the photo below. A nice feature on some high-quality nailers is a built-in depth adjustment fitted into the nose, like the one shown in the bottom photo. A turn of the knob or a thumbwheel is all it takes to adjust the depth to the desired setting. ▼

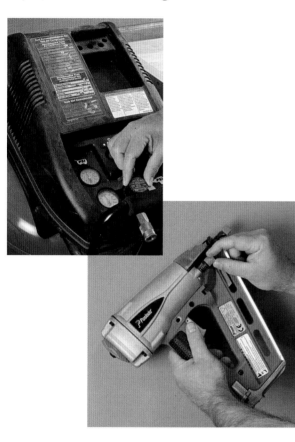

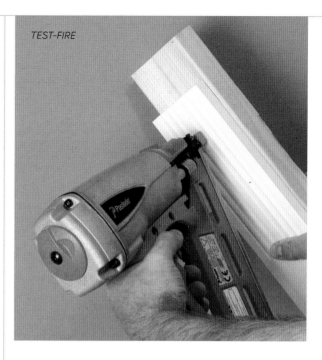

TEST-FIRE

Before you use a nailer to attach your trim, it's a good idea to test-fire a fastener into a scrap of the trim, as shown in the top right photo, to make sure the depth adjustment is correct. Just hold the trim in place against a scrap 2x4, press the nosepiece into the trim, and pull the trigger (make sure your hands are a safe distance away from the fastener). Adjust as needed.

SAFETY FIRST: AIR TOOL SAFETY

1. Wear eye and hearing protection.

2. Always disconnect the air hose when loading or removing fasteners or servicing the tool.

3. Make sure the compressor is supplying the recommended pressure to the tool; do not exceed the maximum pressure of the tool.

4. Never carry an air nailer with your finger on the trigger.

5. Never point an air nailer at yourself or at another person.

6. Make sure you have plenty of hose clearance to avoid tripping.

7. Use only the fasteners recommended by the tool maker; it's best to use their brand.

8. Inspect and lubricate the tool periodically, according the manufacturer's instructions.

9. Keep your fingers away from the tip of the gun; wood can make fasteners deflect, causing an injury.

10. Make sure there's no one behind your work area. Air tools are powerful enough to shoot fasteners through some materials.

77

TROUBLESHOOTING AIR TOOLS

The most common problem you'll encounter with an air nailer is a fastener jam. Jams can often be prevented with routine maintenance, as described in the sidebar on the opposite page. Other nailing problems include fasteners that protrude and those that "stair-step."

FASTENER JAMS

Fastener jams can be frustrating, but are usually easy to clear as described below. Common causes of jams include: a dirty nailer or fasteners, improper air pressure, lack of lubrication, and using the wrong brand or type of fastener. We recommend that you use the same brand fastener as your nailer.

Remove fasteners. To clear a jam, start by removing the fasteners from the nailer, as shown. This will prevent the magazine follower from applying tension to the jammed fastener. ▼

Access the fastener. The next step is to access the jammed fastener. On many nailers, like the one shown here, this is simply a matter of releasing a cover latch to expose the end of the nailer. ▼

Clear the jam. Then reach in with a pair of pliers as shown and gently remove the fastener. Check to make sure that there is no debris in the tip; see your owner's manual for specific clearing instructions. ▼

FASTENERS PROTRUDE

Fasteners that protrude like the one shown below are most often caused by insufficient air pressure or improper depth adjustment—both are easily remedied with a simple adjustment. ▼

FASTENERS STAIR-STEP

Stair-stepped nails or "staircasing" is a common problem with air nailers. It's a sure sign that the compressor that's powering your nailer doesn't have enough punch. Basically, the nailer isn't getting sufficient air to drive nails to a consistent depth, as shown. A more common example of this is when the nails stand proud of the workpiece: The first fastener will be driven all the way in, the next will be proud a bit, and the next even more. The best solution is to use a beefier compressor. If this isn't possible, slow down your nailing to let the compressor fill the tank to sufficient pressure. ▼

PRO-TIP: ROUTINE MAINTENANCE

Empty the tank. When air is compressed inside a tank, moisture in the air condenses to form water. If you don't remove this from the tank, it can rust and eventually rupture. All compressors have a petcock on the underside of its tank so that you can prevent this from happening. Unplug the compressor and loosen the petcock to let pressurized air in the tank blow out water, as shown in the top right photo. After 5 or 10 seconds, close the petcock. ▼

Clean and lubricate tools. Keep the nosepiece and drive pin area clean to help prevent jams. Also, most air nailers require lubrication to run smoothly. Before using any lubrication, check your owner's manual to make sure your nailer requires it. Some nailers are "oil-free" and require no internal lubrication. ▼

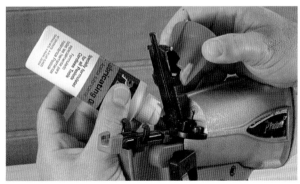

FINE-TUNING TRIM

Even with accurate cuts, odds are that you'll often need to fine-tune a piece of trim to get a perfect fit. This is mainly due to unevenness in walls, as well as working with walls, ceilings, and floors that aren't level or plumb.

TRIMMING MITERS

Miters often require attention, since few corners are truly 90 degrees. This often leads to confusion as to what part to fine-tune to get a good fit. As a general rule, you'll want to focus your attention on one part of the miter, as illustrated below. ▼

ANGLES LESS THAN 45°

ANGLES MORE THAN 45°

TRIMMING WITH A BLOCK PLANE

A favorite tool of pros for fine-tuning trim is a block plane like the one shown in the top right photo. As long as the blade is sharp, you can fine-tune a piece of trim with a few quick strokes. Make sure to hold the workpiece firmly and skew the plane to produce more of a shearing cut. Work with the grain whenever possible, and if you must plane against the grain, take precautions to prevent chip-out. Either clamp a scrap block to the edge of the wood, or plane in toward the center from both directions. ▶

WITH A FILE OR RASP

For really fine adjustments on trim, a rasp or fine mill file is invaluable. A four-in-hand rasp like the one shown below combines four tools in one. It features a flat and curved rasp, and a flat and curved file—and its small size makes it the perfect addition to a tool belt. To use a file or rasp, hold the workpiece firmly and use full, smooth strokes. Most files are designed to cut in only one direction, so lift the file at the end of the cut before starting the next stroke. ▼

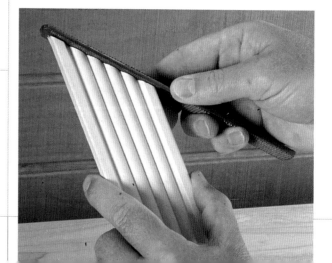

WITH A SHOOTING BOARD

Planing a perfect edge on the end of trim is a challenge. That's because the surface of the trim itself is used as the reference for square, and odds are that it isn't. That's where a shooting board comes in. A shooting board is a shop-made jig that lets you hold the trim square so that you can accurately shave it with a plane. A shooting board has one or more cleats secured to a base at the desired angle. The trim is held against a cleat, and the plane is laid on its side and passed back and forth over the end of the trim to shave it to size. ▼

WITH A POWER SAW

You can fine-tune trim on a power miter saw by adjusting the angle of the blade just a hair and making a cut. Alternatively, you can shim the workpiece with a playing card or two as shown below and then make a cut. ▼

WITH A HAND SAW AND MITER BOX

The same playing card trick you use with a power miter saw can also be used to fine-tune a piece of trim cut with a hand saw and a miter box. ▼

TOUCH-UP WORK

Every trim job will require some touch-up work to mask nail holes and the occasional ding. Putty and wax crayons are used for this. Gaps between trim and adjacent surfaces are filled with caulk.

FILLING NAIL HOLES

How you fill nail holes and what you use to fill them depends on whether the trim is finished or not. Holes in pre-finished trim can be filled with wax crayons (as described below) or with non-hardening putty (as described on page 84). Nail holes in unfinished trim are filled with hardening putty. See the sidebar on the opposite page for more on types of fillers used to hide nail holes. ▼

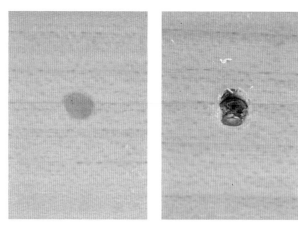

USING WAX CRAYONS

Trim carpenters often pre-finish solid-wood trim with a clear topcoat before installing it. Then they use wax crayons to fill the nail holes once the trim is in place, as described at right.

Apply the wax. To use a wax crayon, start by matching a crayon to the trim. Don't go by the name on the crayon (such as natural oak). Instead, buy a range of colors and return the ones you don't use. Fill a nail hole by scrubbing the tip of the crayon over the hole. Overfill it slightly. ▼

Remove the excess. Now use a clean, dry cloth to buff away any excess wax until it's flush with the surface of the trim. ▼

82

FILLING NAIL HOLES WITH PUTTY

There are two types of putty for filling nail holes: hardening and non-hardening. To apply hardening putty, press it into the nail hole with a putty knife (bottom photo) or a finger (top photo). Leave the putty a bit proud of the surface: You'll sand it flush, as described on page 84. Non-hardening putty is applied in a similar manner as wax crayons, as described on the opposite page. ▼

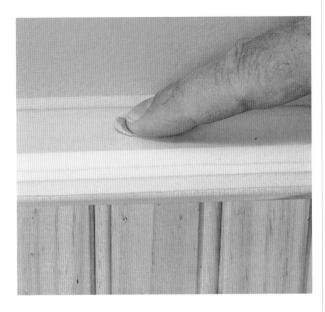

PRO-TIP: TYPES OF FILLERS

Not all fillers are the same. You can use either putty or wax crayons to hide nail holes. There are two types of putty available: hardening and non-hardening. Hardening putty is applied prior to finishing and is typically sanded smooth. Non-hardening putty is color-matched to the finish on your trim (usually natural or stained wood) and is applied after the finish has been applied—it does not harden over time. Like non-hardening putty, wax crayons are applied to finished trim and excess wax is buffed away with a clean, dry cloth. ▼

REMOVING EXCESS NON-HARDENING PUTTY

Excess non-hardening putty is removed just like excess wax. You simply buff away the excess with a clean, dry cloth, as shown. For best results, fold the cloth into a pad to prevent it from dishing out the putty as you buff. ▼

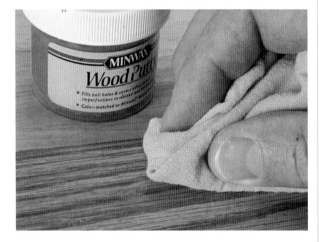

TOUCH-UP WORK

If you've filled nail holes on pre-finished trim with hardening putty and had to sand to get it flush, odds are that you sanded into the finish as well, and it will need to be touched up with paint or with a clear finish. Use light strokes when applying the finish, and feather it away from the filled hole to achieve a smooth transition. ▼

REMOVING EXCESS HARD PUTTY

To remove excess hardening putty, first let the putty dry the recommended time. Then sand it flush with the surface. Wrap a piece of open-coat sandpaper around a scrap block, as shown. The block prevents you from dishing into the putty as you sand. ▼

CAULKING INTERIOR TRIM

Since walls, ceilings, and floors are often uneven, you'll end up with gaps between trim and adjacent surfaces. On interior trim that will be painted, these gaps can be filled with a paintable latex caulk (as shown in the top right photo), then painted to match the trim. With solid trim finished with a clear coat, you can either use clear silicone caulk or find a tinted caulk that matches the trim to conceal any gaps.

TYPES OF CAULK

You'll find it handy to have three types of caulk on hand for most jobs: acrylic latex, silicone, and siliconized acrylic latex.

Material	Application
Acrylic latex	Filling gaps and voids behind interior trim; also known as painter's caulk, as it accepts paint well; inexpensive and very easy to use and clean up
Silicone	Filling gaps and voids in exterior trim; extremely flexible and long-lasting, but can't be painted
Siliconized acrylic latex	Interior and exterior applications where a more flexible caulk is needed; bonds better to surfaces and lasts longer than acrylic latex; less messy and easier to use than silicone

CAULKING INTERIOR TRIM

CAULKING EXTERIOR TRIM

Gaps between exterior trim and adjacent surfaces should always be filled with 100% silicone caulk, as shown below. Silicone caulk will seal the perimeter to protect against drafts while still remaining flexible over time. This way it can expand and contract as the trim does with seasonal changes in humidity. ▼

MAKING YOUR OWN TRIM

There are many reasons why you may want to make your own trim: You're trying to match existing trim and can't find a match, or you just want to be creative. If you've got a couple of woodworking tools—a router and a table saw—it's easy to make your own. In addition to simple routed edge profiles, you can also create complex profiles by gluing up profiled strips, as shown here. ▼

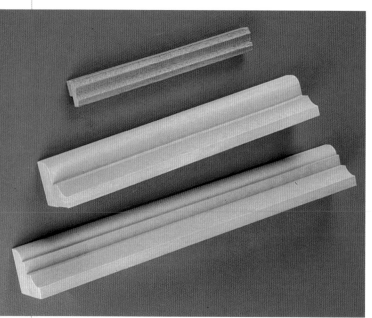

ROUTER BIT PROFILES

Your main tool for creating custom trim is a router (page 36) fitted with a variety of router bits. Since there are dozens of profiles available (as illustrated below), the variety of trim profiles you can create are almost endless—especially if you create built-up trim as described on page 88. ▼

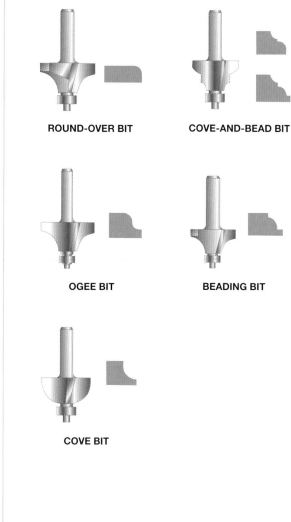

ROUND-OVER BIT   **COVE-AND-BEAD BIT**

OGEE BIT   **BEADING BIT**

COVE BIT

In addition to built-up trim (page 88), you can also create your own trim from a single piece. This may entail routing decorative edges, cutting coves, and/or bevel-ripping the backs.

ROUTING DECORATIVE EDGES

The only challenge to working with a decorative profile bit, like a Roman ogee, for example, is not

cutting the full profile in a single pass. Beginners often lower the bit to expose the desired profile and start routing. This will cause several problems, including chip-out, burning, and excessive wear and tear on both the bit and the router. Take light cuts and sneak up on the final profile.

CUTTING COVES

CUTTING COVES

Cutting coves on the table saw is done by angling an auxiliary rip fence so it's not parallel to the blade. Angling the fence like this will produce a scooped cut. To determine the angle for the auxiliary fence, first mark both the width and depth of the cove to be cut on a scrap cut to the same dimensions as your finished trim. Now clamp an auxiliary fence (just a straight piece of scrap lumber) to the saw top so that it's aligned to cut the cove—this will take some trial and error, as shown in the top right photo. As you'll be pushing the trim against this fence, make sure it's clamped firmly in place. And since you'll be presenting the trim to the blade in a manner that the blade wasn't optimized for, you'll want to take many light passes, as shown in the middle right photo. It's best to take no more than $1/16$" off in a single pass.

BEVEL-RIPPING BACKS

If you want bevels cut on the back of your trim so that it can be installed at an angle, all you need to do is tilt the blade on your table saw to the desired angle, set the rip fence for the desired width of cut, and push the trim past the blade. ▼

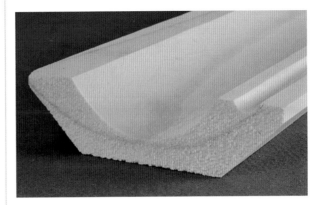

PRO-TIP: BUILT-UP TRIM

If you need a profile and can't find a single router bit to do the job, consider breaking the profile down into simpler sections. Then you can rout profiles on stock of varying thickness, and stack or build up the strips to form the desired profile, as illustrated below. ▼

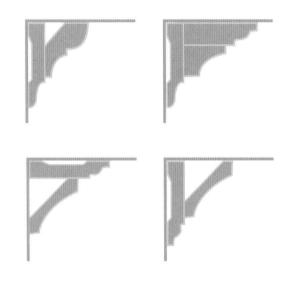

ROUT THE PROFILES

Assemble the strips. With all the profiles routed, simply glue the strips together with the back edges flush. Spring clamps, like those shown below, will apply sufficient clamping pressure. Alternatively, strips of inner tube can be stretched around the profiled strips to hold them together until the glue dries. ▼

Rout the profiles. To make your own trim, begin with a wide piece of wood that's cut a bit longer than your finished trim length. Then use a router fitted with the desired profile to rout a profile along its full length as shown in the top right photo. Although the full profile is shown here, it's best to take multiple passes—usually three light cuts—to sneak up on the final profile. This is especially important when working with thin stock. Once you've profiled the edge, use a power saw to rip the trim to the desired width. A table saw is the best tool for this job, but you can also cut the trim with a circular saw fitted with a rip fence.

REMOVING TRIM

Trim jobs often begin with removing old trim. How you remove the trim will depend on whether or not you plan to reuse it. Either way, it's always best to use a putty knife first to cut any caulk used to conceal gaps. When dry, caulk can form a strong bond. If you don't cut the caulk before pulling off the trim, you'll probably pull off part of the wall covering with the trim.

FOR DISPOSAL

A cat's paw is a special prybar designed for removing nails flush or below the surface of a workpiece. It can make quick work of removing trim that won't be reused. The sharp tip of the claw is driven into the wood surrounding the nail with a hammer, as shown below. Then the tool is leveraged to pull out the fastener. A cat's paw is a destructive tool, as it cuts into the surface to gain a purchase on the nail; so it should be used only to remove trim for disposal. ▼

FOR REUSE

To remove trim for reuse and to prevent a prybar from damaging the wall, slip a putty knife between the trim and your prybar, as shown. Now gently pry the trim away from the wall. To remove the nails, see the sidebar below. ▶

PRO-TIP: REMOVING NAILS

If you're planning on reusing any of the trim that you've removed, don't pound the nails out through the face of the molding. This will split the wood, creating a larger hole to fill. Instead, pull the nails out from behind with a pair of locking pliers or end nippers, as shown below. This creates a much smaller hole to fill. ▼

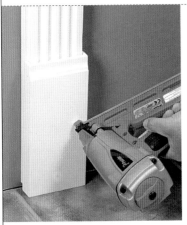

CHAPTER 4:

FLOOR AND WALL TRIM

The most noticeable trim in a room is often attached to a wall, like the wall frames in the dining room in the photo on the opposite page. Other examples of highly visible wall trim include: baseboard, chair rail, wainscoting, plate rail, and pilasters. In this chapter we'll show you how to install each of these and include tips and techniques to help you achieve professional-looking results.

ONE-PIECE BASEBOARD

One-piece baseboard is the simplest baseboard to install. Trim manufacturers offer a wide variety of profiles and sizes. Profiles range from a simple, single roundover on top to complex, sculpted shapes that can run the full width (height) of the baseboard. Sizes vary from 2½" up to over 6" in width (height). Most baseboard comes in 8- or 16-foot lengths. As a general rule of thumb, use wider (taller) baseboard in rooms with higher ceilings. Also, if your floors are uneven and you can't level them, consider using foam trim—it's more flexible than wood and can be gently shaped to hug irregularities in the floor.

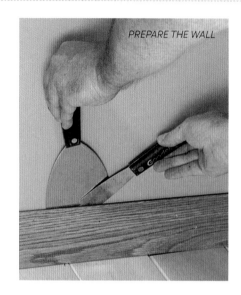

PREPARE THE WALL

PREPARE THE WALL

Before you start cutting trim, you'll first have to remove the old trim, as shown in the top right photo and described on page 89. It's also important to note that the new baseboard will lie flat on a wall only if the wall is flat and plumb. Trim carpenters know that drywall joint compound can build up at joints (especially at corners) and ruin the fit. That's why you'll often see them pull out a putty knife and scrape a section of a wall flat. This is much more efficient (and better-looking) than installing trim on an uneven wall and filling in the resulting gap or gaps with caulk.

LOCATE AND MARK STUDS

With the wall prepped, the next step is to locate the wall studs with an electronic stud finder and mark their position. ▼

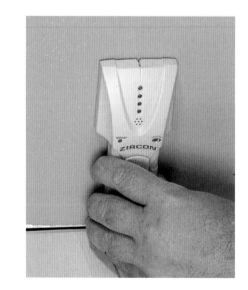

START IN ONE CORNER

To install baseboard, measure the longest, most visible wall and cut a piece of baseboard to fit; butt it into the corner as shown in the photo below. The typical sequence used to install perimeter trim is illustrated here. ▼

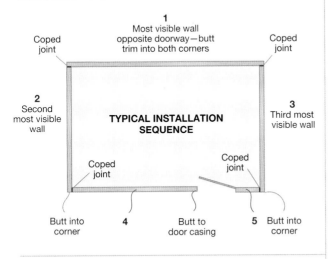

1
Most visible wall
opposite doorway—butt
trim into both corners

Coped joint

Coped joint

2
Second most visible wall

TYPICAL INSTALLATION SEQUENCE

3
Third most visible wall

Coped joint

Coped joint

Butt into corner

4

Butt to door casing

5 Butt into corner

ATTACHING BASEBOARD

Baseboard is machined with a recess on its back, as illustrated in the drawing at right. This creates two nailing flats, which helps the trim lie flat on uneven walls. Trim carpenters usually secure baseboard in two places: to the sill or sole plate, and to the studs, as shown in the bottom photo. Notice that this is where the nailing flats are located. If you drive a fastener into the middle of the trim—especially with a hammer followed by a nail set—you risk cracking the trim since it's unsupported here. Pros often use a heavier-gauge nail at the sill plate (typically 16-gauge), and a lighter-gauge nail or brad at the more delicate top portion of the trim to prevent it from splitting. ▼

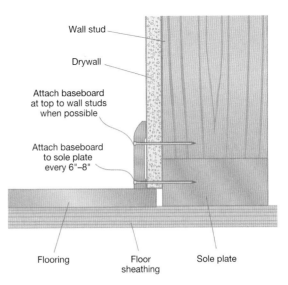

Wall stud

Drywall

Attach baseboard at top to wall studs when possible

Attach baseboard to sole plate every 6"–8"

Flooring

Floor sheathing

Sole plate

COPE THE NEXT PIECE

Now you can move on to the next piece of baseboard (as illustrated in the sequence drawing on page 93), coping one end as described on pages 70–71 and shown below. ▼

SCARF IF NEEDED

For walls that are longer than your baseboard, you'll need to splice together pieces with a scarf joint, as shown below. On a scarf joint, the ends of the connecting pieces are mitered at opposing angles to mesh with one another. This creates a nearly invisible seam. ▼

MITERED RETURN

Occasionally, baseboard needs to stop before reaching a corner. Instead of cutting a butt or angled miter and leaving the end grain exposed, pros make a mitered return. This way no end grain is left exposed, and the trim is neatly terminated. To make a mitered return, begin by back-mitering the trim piece to leave a 45-degree end that faces the wall. Now miter-cut a scrap of trim at 45 degrees opposite that of the back-cut miter that you just made. Hold this end up against the molding and mark it to length. Then use a hand saw to cut the return to length so the 90-degree end butts up against the wall. Apply glue to the mitered ends of the trim and return, and press the pieces together. Apply masking tape across the joint to hold the pieces in place until the glue dries. ▼

CONCEAL THE NAIL HOLES

With all the baseboard in place, fill in all nail holes as described on pages 82–84. How you do this and what filler you use will depend on whether or not your trim is pre-finished. ▼

FILL ANY GAPS

Even if you fine-tune the fit of the baseboard, chances are that you'll still end up with some gaps between the top of the baseboard and the wall covering. Fill these gaps with paintable latex caulk or caulk that's color-matched to the trim, as shown below. Wipe off any excess immediately, and when dry, paint if necessary. ▼

PRO-TIP: CORNER BLOCKS

If you don't want to cope or miter your baseboard to fit the inside and outside corners of the room, consider installing pre-made corner blocks. These profiled blocks are glued and nailed into the corners. The baseboard is cut straight on its ends and simply butts up against the corner blocks. ▼

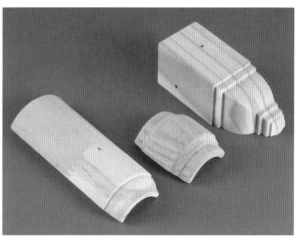

PRO-TIP: BUILT-UP BASEBOARD

Built-up baseboard is just multiple pieces of trim stacked on top of each other. It can be as simple as a base molding and a base cap or as complex as a three- or four-layer buildup, as illustrated below. Naturally, the more layers, the more exotic the profile. Besides adding visual interest, there are two other reasons for using built-up baseboard. First, it can actually be less expensive to purchase simple trim and stack it up than to buy single heavily profiled trim. Second, if you're matching trim to an existing look, the trim may not be available. By mixing and matching stock trim, you can often create the look that you're after. If you still can't match it this way, consider making your own trim (see pages 86–87). ▼

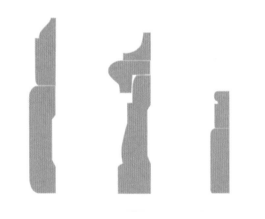

Install the base trim. To install built-up baseboard, locate the wall studs and mark their position. Then measure the longest, most visible wall, cut a piece of base to fit, and butt it into the corners. Attach the base as shown in the top right photo, and repeat for the remaining base pieces as you work your way around the room.

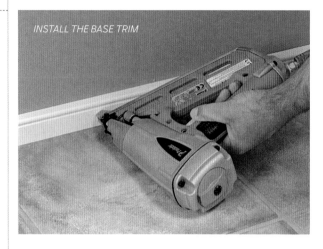

INSTALL THE BASE TRIM

Attach additional layers. Now install the next piece (or base cap if you're using a two-piece trim buildup, as here) over the base trim. Some base cap is rabbeted on its bottom inside face to fit over the base trim. Others (like the base cap shown here) do not have a rabbet. Cut and fit the base cap as needed, and secure it to the wall studs or base trim. You'll want to use a lighter-gauge fastener here to prevent splitting. Repeat for all the base cap. ▼

PRO-TIP: OUTSIDE CORNERS

Outside corners for baseboard are mitered. The only real challenge here is determining the angle of the corner, as it most likely isn't 90 degrees. See pages 59–60 for more on measuring angles and page 66 for information on cutting miters.

Cut and install the first half. Measure the corner angle and divide this result by two. Cut a piece of baseboard to this angle and attach it to the wall, taking care to align the inside edge with the exact corner of the wall. ▼

Cut and install the second half. Cut the second piece of baseboard and check the fit. Fine-tune the fit as described on pages 80–81 to get the tightest possible miter. Then secure it to the wall as shown in the top right photo.

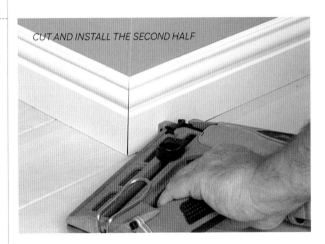

CUT AND INSTALL THE SECOND HALF

Pin the miter. Even if there are no gaps at the miter joint, it's a good idea to pin the miter joint as shown below. This prevents the miter from opening up as the trim adjusts to seasonal changes in humidity. ▼

INSTALLING CHAIR RAIL

Chair rail was originally created to protect walls from getting dinged up by chairs. But chair rail can also be decorative: It's often used to create a border between top and bottom wall expanses. Many homeowners like to dress the walls above and below the chair rail differently. So, you could paint the sections of the wall different colors, or wallpaper one section and paint the other. If you are planning on painting or wallpapering the room where you'll be installing the chair rail, do so now before installing the chair rail. You can purchase pre-made chair rail in a variety of shapes, or you can make your own profile by building up the molding using several pieces, as described on pages 86–87.

LOCATE THE RAIL

Chair rail is typically installed 32" to 36" above the floor, as illustrated in the drawing on the opposite page. Measure this distance around the room and mark it by snapping a chalk line, as shown in the middle photo. Alternatively, you can use a laser level as described below. ▼

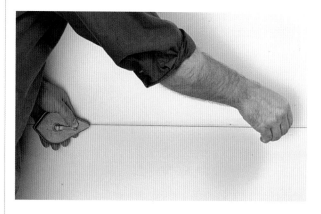

PRO-TIP: LASER LEVELS

You could snap a chalk line around the perimeter of your room to locate the chair rail, but then you'd have to remove the chalk from your walls. A less messy alternative is to use a laser level, like the one shown here. The advantage of a laser level is that it shoots a perfectly level line along a wall—or around the perimeter of a room—without leaving any marks. ▶

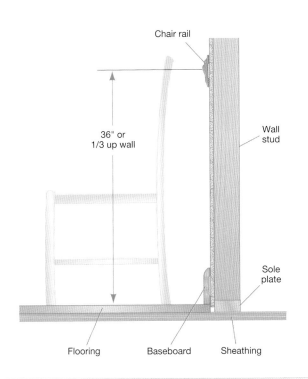

Chair rail

36" or
1/3 up wall

Wall stud

Sole plate

Flooring   Baseboard   Sheathing

LOCATE AND MARK THE WALL STUDS

After you've marked your reference line for the chair rail, use a stud finder to locate the wall studs in the room, as shown in the photo below. Use a pencil to mark each location. This is where you'll secure the chair rail to the wall. ▼

INSTALL THE RAILING

Cut your first piece of chair rail and position it so its bottom edge aligns with the chalk line you snapped earlier (use the same installation sequence as described for baseboard on page 93). A helper to hold the long strips of chair rail will make this job go a lot quicker. When aligned, secure the chair rail to the wall at each stud location with 2"-long finish nails. If you're running the chair rail around the room, you'll need to either miter or cope the ends to fit into the corners. When all the chair rail is in place, go back and apply putty to conceal any nail holes. ▼

WALL FRAMES

Need to break up a large expanse of wall or add visual interest to an otherwise dull room? Consider installing wall frames. Wall frames can be large or small, and installed with or without a chair rail or other molding. The frames can be oriented vertically or horizontally and can be installed on the upper and/or lower portion of a wall. See the sidebar below on how to size your wall frames.

LOCATE THE FRAME TOPS

After you've sized your frames as described in the sidebar at right, you'll need to locate them on the wall. Where you locate them is a matter of personal preference. Some like to center the frames on a wall; others do not. The simplest way to locate them is to use masking tape to mock up how they'd look on a wall. This way you can experiment easily until you find a location that pleases your eye. Once you've determined the frames' location, use a laser level (as shown in the top right photo) or snap a chalk line to serve as a reference point for installing the frames. Then locate and mark the wall stud locations along this reference line. Now mark where your frames will be installed.

LOCATE THE FRAME TOPS

PRO-TIP: GOLDEN RECTANGLE

Designers, architects, and engineers the world over have been using the "golden ratio" for centuries. Euclid defined this proportion that has uniformly been accepted as the one most pleasing to the eye. (His complicated formula involved dividing a line into what Euclid called its "extreme and mean ratio." This ratio is equal to the ratio of 1 to 1.618.) The amazingly consistent occurrence of this ratio in nature has led scientists, mathematicians, artists, and engineers worldwide to call it the "divine proportion" or "golden ratio." When used to construct a rectangle, this ratio produces what's called a golden rectangle—the perfect solution for sizing your wall frames. ▼

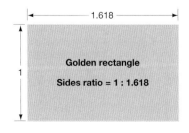

1.618

1

Golden rectangle

Sides ratio = 1 : 1.618

BUILD THE FRAMES

Cut your frame pieces to size and then make a simple jig to assemble them. The jig is just a pair of rectangles: a form and a base. The form is cut to the exact inner dimensions of the frame and attaches to the base. This automatically squares up the frame as it's assembled. Apply adhesive to the mitered ends, place the frame parts on the jig, and pin the ends together with brads. ▼

APPLY AN ADHESIVE

Odds are that many of the frames will not lie on top of a wall stud. This means you'll need to rely on an adhesive to attach them securely to the wall. Apply adhesive to the back of the frame, as shown in the top right photo.

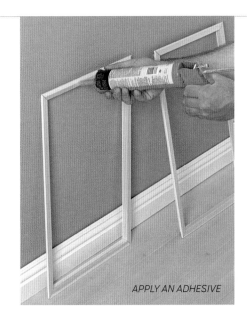

APPLY AN ADHESIVE

ATTACH THE FRAMES TO THE WALLS

Now you can align the top edge of the frame with the chalk or laser line and your frame layout marks. Press the frame into the wall to distribute the adhesive evenly, and secure the frame to a wall stud if possible. Repeat for the remaining frames. Fill all nail holes, and apply a finish if you didn't pre-finish the trim before installation (which we recommend). ▼

INSTALLING WAINSCOTING

Wainscoting protects walls and gives a room a rich finishing touch. Most wainscoting is vertical boards attached to the lower half of a wall. (Another version, known as frame-and-panel wainscoting, is described on pages 110–115.) On tongue-and-groove wainscoting (shown here), the edges of the boards are milled with a tongue on one edge and a groove on the other to join the boards together, as illustrated in the drawing below left. Additional trim handles the transitions between the wainscoting and the wall and floor. These include base trim, a cap, and an optional scotia piece that fits under the cap. See the sidebar on page 105 on how to condition wainscoting before you begin your installation.

To prepare a room for wainscoting, start by removing any receptacle and switch covers that will be affected. Then remove the baseboard around the perimeter of the room as described on page 89. Next, you'll need to create a nailing surface for the boards, as described on page 104.

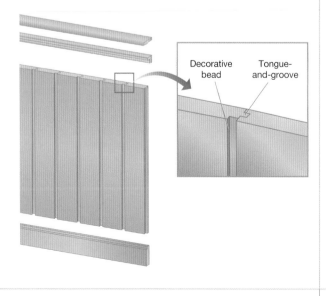

Decorative bead   Tongue-and-groove

LOCATE THE WAINSCOTING

To install tongue-and-groove wainscoting, begin by locating the top of the wainscoting. Measure the length of your wainscoting, and transfer this measurement directly to the wall. ▶

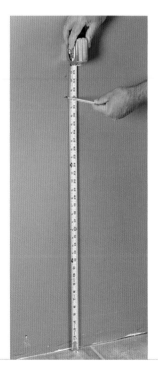

SNAP A CHALK LINE

Now you can snap a chalk line at the top of the wainscoting. Take care to make this a level line, and snap it wherever you'll be installing wainscoting. The wall covering below this line will be removed in the next step to install a backer panel, as described on page 104. If you'll be using backer strips, snap a series of horizontal lines to identify the drywall that you'll remove for the strips. ▼

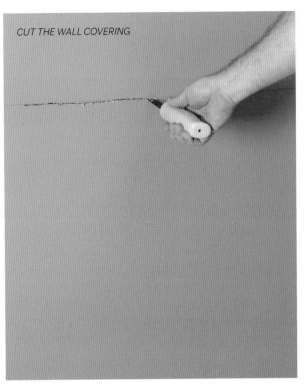

CUT THE WALL COVERING

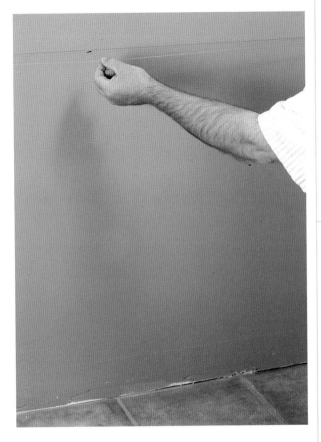

REMOVE THE WALL COVERING

Once cut, you can pull the wall covering away from the wall as shown and discard it. Remove any fasteners remaining in the wall studs, and vacuum away any drywall dust or debris. ▼

CUT THE WALL COVERING

Now you can cut through the wall covering. In most cases, this means using a drywall saw, as shown in the top right photo. If you don't have a drywall saw, you can use a utility knife—it'll just take a series of cuts to break all the way through. Even with a drywall saw, you'll need to cut through the drywall at the stud locations with a utility knife.

PRO-TIP: BACKER STRIPS AND PANELS

The milled vertical edges of tongue-and-groove wainscoting slip together to create a continuous panel. To keep this "panel" in place, each board needs to be attached firmly to a wall. The problem is that wall studs are spaced 16" apart on center. This means that you'll need to create some type of nailing surface along the full width of each wall for the wainscoting. The two most common ways to do this are with backer strips or with backer panels. To make up for the thickness of the strips or panel, the wall covering is removed to allow the wainscoting to be as flush to the wall as possible. (Note: If you don't mind wainscoting that extends out from the wall, you can install backer strips or panels without first removing the wall covering.)

Backer strips. Backer strips are horizontal strips of wood (typically 1x2's) that are attached to the wall studs. To allow the wainscoting to sit as flush as possible to the wall covering, the wall covering is cut away. ▼

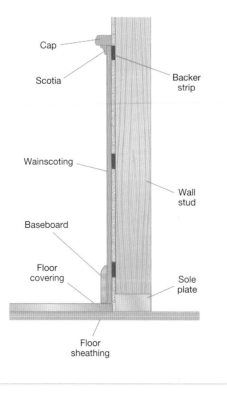

Backer panel. A less messy alternative to cutting grooves in drywall is to simply remove the drywall below the wainscoting and install a continuous plywood backer. ▼

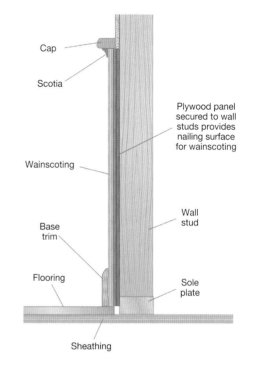

SECURE THE BACKER PANEL TO THE STUDS

Measure and cut a piece of plywood to fit in the opening you made in the wall covering. For $\frac{1}{2}$" drywall, use $\frac{1}{2}$" AC plywood and make sure the good side faces out. Attach the panel to the wall studs every 6" to 8" along the length of the studs. ▼

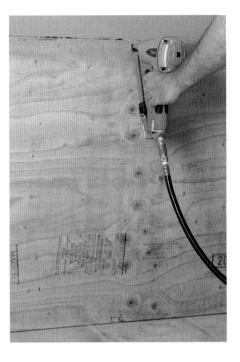

INSTALL THE FIRST PIECE

The first piece of wainscoting that you install is critical, since all other pieces reference off it. So make sure to use a level when installing it so that it goes in plumb, as shown. Secure the piece as illustrated in the drawing below. ▶

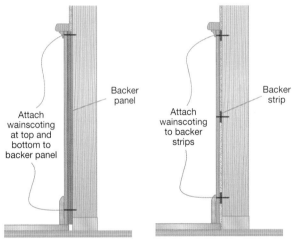

Attach wainscoting at top and bottom to backer panel

Backer panel

BACKER PANEL

Attach wainscoting to backer strips

Backer strip

BACKER STRIPS

PRO-TIP: CONDITIONING WAINSCOTING

Solid wood expands and contracts as it reacts to humidity—and solid wood wainscoting is no exception. That's why it's important to "condition" your wainscoting prior to installation. This involves "stickering" the wainscoting in the room where it will be installed so it can acclimate to the room's humidity. Set the pieces on stickers (scraps of $\frac{3}{4}$"-square dry wood) to let air circulate freely. Most manufacturers suggest allowing the wainscoting to acclimatize for at least two days before installation. If you don't do this, the pieces can shrink or swell once they're installed, creating buckling or gaps.

CONTINUE ADDING BOARDS

With the first piece of wainscoting installed, it's easy going. Just mate the tongue-and-groove edges of additional pieces together and secure them to the backer panel or strips. It's a good idea to stop every three or four pieces and recheck with a level to make sure they're still plumb. If they're not, just shift the next piece slightly to bring it back into plumb. Continue working your way around the room, cutting pieces to fit at the end of each wall as needed. See the sidebar for typical ways to treat corners. ▼

PRO-TIP: INSIDE AND OUTSIDE CORNERS

You can easily turn inside and outside corners with wainscoting. On inside corners, simply butt the pieces up against each other, as illustrated in the drawing below. Odds are that you'll need to scribe one or both pieces to fit tight against each other, as described on pages 56–57. Outside corners can also be butted up against each other, as illustrated in the bottom drawing. The only problem here is dealing with the tongue or groove on the edge of the boards. One way to handle this is to simply conceal them with matching corner trim, as shown. Alternatively, you can cut off either the tongue or the groove and butt the parts together. ▼

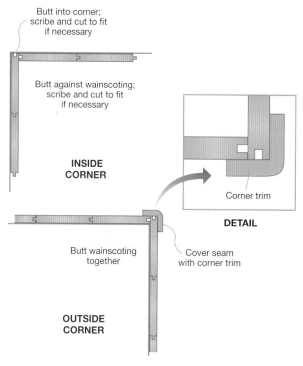

Butt into corner; scribe and cut to fit if necessary

Butt against wainscoting; scribe and cut to fit if necessary

INSIDE CORNER

Corner trim

DETAIL

Butt wainscoting together

Cover seam with corner trim

OUTSIDE CORNER

INSTALL THE BASE TRIM

When all the wainscoting is installed, you can install the base trim. Cut it to length and secure it to the wainscoting and wall studs. Cope the ends as necessary, as described on pages 70–71. ▼

INSTALL THE CAP

INSTALL THE CAP

The cap or nosing rests on top of the wainscoting and is held in place with nails. Cut pieces to fit and secure them to the top backer panel or strip, as shown in the top right photo. On both inside and outside corners, join the cap with miter joints.

INSTALL THE SCOTIA

Gaps between the cap and the wainscoting can be concealed with scotia. Scotia also creates a smoother transition between the cap and wainscoting. The scotia fits under the cap and is typically $\frac{1}{2}$" or $\frac{3}{4}$" cove molding. When done, fill all nail holes and apply the finish of your choice. ▼

PRO-TIP: QUICK-INSTALL WAINSCOTING

If you like the idea of wainscoting but don't want to hassle with removing drywall and installing backer strips or panels (as described on page 104), consider using quick-install wainscoting. Quick-install wainscoting "systems" utilize fit-together panels that are machined on the ends to fit into matching base and cap moldings, as illustrated in the drawing at right. This makes installing wainscoting a simple process that can easily be accomplished in a weekend. To prepare a room for wainscoting, start by removing any receptacle and switch covers that will be affected by the wainscoting. Then, using a prybar and wide-blade putty knife to protect the wall, pry off the baseboard molding around the perimeter of the room.

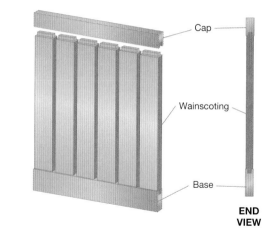

Cap

Wainscoting

Base

END VIEW

Locate the wall studs. The next step is to locate the wall studs. These will be used to nail in place any panels that sit over a stud, but more importantly, to attach the base and cap moldings securely to the wall. Use an electronic stud finder to locate these, and then mark each location with a pencil. ▼

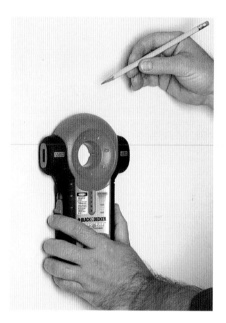

Install the base. With the room prepped, the next step is to attach the base to the wall at each of the marked stud locations. Check this with a level as you install it to make sure that it creates a level foundation for the wainscoting panels. Cut the base as needed when you encounter corners, and work your way around the room. ▼

Install the wainscoting. With the base in place, you can add the panels. With most systems, these simply slip into a groove in the top edge of the base. Since the panels are tongue-and-grooved together and are also held in place with the cap molding, you don't need to affix each piece to the wall. Occasionally check the panels with a small level to make sure they're going in plumb, as shown in the top right photo; adjust as necessary. Keep adding panels, attaching them to the wall whenever the panel sits over a wall stud. Drive a nail through the panel and into the stud. If you encounter any electrical receptacles or switches, mark and cut the panels as needed to wrap around the electrical box.

INSTALL THE WAINSCOTING

Add the cap. When all the wainscoting is in place, add the cap. The bottom edge of this should be grooved to fit over the panels. Set the cap in place and check it with a level before securing it to the wall studs with nails. Install the remaining cap and then go around the room and fill in any nail holes with putty; sand flush when dry. ▼

FRAME-AND-PANEL WAINSCOTING

There are many types of wainscoting, but frame-and-panel is the most formal-looking. That's because it consists of panels set into frames, as shown in the photo below. In the past, the panels fit into grooves cut in the frames. This took considerable woodworking skill. A simpler version is to attach a frame to a solid back panel, as illustrated in the drawing at right. The frames of yesteryear were joined together with stout mortise-and-tenon joints, which also required considerable skill. Fortunately, modern technology—this time in the form of a specialized router bit—makes it easy to join together the frames. Alternatively, you can join together frames by butting them together and strengthening the joints with biscuits or dowels.

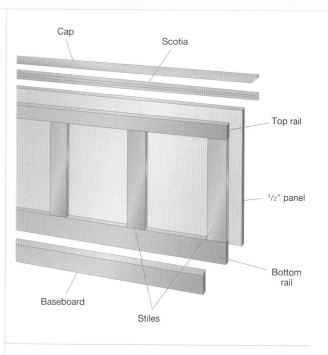

Cap

Scotia

Top rail

1/2" panel

Bottom rail

Baseboard

Stiles

RAISED PANELS WITH A ROUTER

If you're after the raised panel look for your wainscoting, or if you've used a cope-and-stick set to join together your frame pieces, you'll need to make some raised panels. These can be made with a table-mounted router (as described here) or on the table saw (as described on page 112). ▼

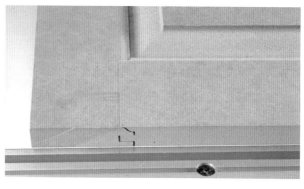

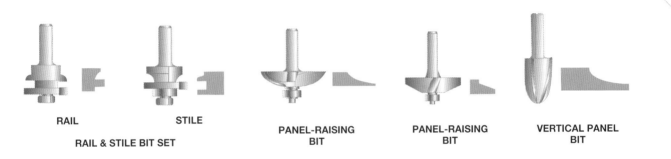

RAIL   **STILE**

RAIL & STILE BIT SET

PANEL-RAISING BIT

PANEL-RAISING BIT

VERTICAL PANEL BIT

ROUTING EDGES

Profiling the edges of a panel does three things: It creates an attractive panel; it reduces the edge thickness so the panel can fit in the grooves of the frame; and it "raises" the center of the panel, creating a richer, more 3-D effect. Panel raising requires special bits. These are either horizontal (not recommended) or vertical, as illustrated in the drawing above. Note: These bits are large and should be used only on routers mounted in a table.

Because vertical panel-raising bits require you to rout the workpiece on its edge, you need to attach a tall auxiliary fence to your existing fence to provide stable support to the workpiece, as shown. Once the fence is in place, adjust the bit for the first of many passes. There are two ways to make multiple, light passes with one of these bits. One way is to slowly raise the bit up between passes until the full profile is achieved. The other method is to adjust the bit height for a full profile and then move the fence forward to take a light cut, easing it back to slowly expose more of the bit until the full profile is cut. Whichever method you choose, make sure to start routing the end grain first. This way, if there's any tear-out, it'll be removed when you rout the edge grain. ▶

PRO-TIP: COPE-AND-STICK FRAMES

The cope-and-stick joint is very strong and attractive, and simplifies construction since it also cuts the groove for the panel at the same time as the joint. Cope-and-stick joints can be routed with separate bits—always sold as matched sets—or with a single, multipurpose bit, as shown here. Read and follow the bit manufacturer's directions regarding setup and use of their cope-and-stick bits.

Rout the stick. The stick half of the joint is routed along the inside edges of the frame rails and stiles of a frame as shown. It accepts the panel and also the coped end of the rails.

Rout the cope. The cope half of the joint is routed on the ends of the rails. Note: Since cope-and-stick bits are large, they should be used only on a table-mounted router. With both parts routed, you can check the fit; see the bottom right photo on page 110 for how these parts fit together along with their panel.

PRO-TIP: **RAISING PANELS ON A TABLE SAW**

There are two methods for raising panels with a table saw: one-pass and two-pass. Because you cut the workpiece on its edge when raising a panel, you need to attach a tall auxiliary fence to your rip fence to provide stable support to the workpiece, as shown in the photo. ▼

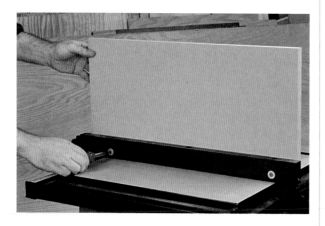

One-pass method. With the one-pass method, you cut the field (the angled part of the panel) and the shoulder (where the field meets the face of the panel) in a single pass. This is simply a matter of tilting the blade to the desired angle and raising it to the desired height. There are two disadvantages to this method: The shoulder of the raised portion can only be as deep as the thickness of your saw blade (typically 1/8"), and the shoulder will be angled, not perpendicular to the face of the panel. That's why we prefer the two-pass method described at right.

Two-pass method. The two-pass method creates the field and the shoulder separately in two passes. Although this takes longer than the one-pass method, it offers some advantages. First, you can cut the field anywhere along the edge of the workpiece. This means that you can vary the thickness of the edges of the panel so they can fit into the grooves cut in the frame pieces. The depth of the shoulder can also be varied depending on where you cut the field. If you want a 1/4"-deep field, you can make it so. Additionally, since the workpiece is laid flat on the saw top, the shoulder will be cut perpendicular to the face of the panel. ▼

Kerf to define field.
The first-pass cut defines the field of the panel. Adjust the rip fence and blade height to cut the desired field. Make a test cut on scrap wood and then cut

your project panels once the setup is correct. Make the end-grain cuts first, then the long-grain cuts to minimize tear-out.

Cut the field. Now adjust the blade to cut the field at the desired angle and position the rip fence to define the thickness of the panel's lip. Make a test cut and when satisfied, cut all your panels, taking care to make the end-grain cuts first to minimize tear-out.

LAY OUT THE FRAMES

The next step is to lay out the frames on the backer panels, as shown below (see page 104 for more on backer panels). You have two options on sizing the panels: You can make them all the same size and live with partial panels on one end of each wall, or you can custom-size panels for each wall so that each panel ends up equal in size. (For more on sizing panels, see the sidebar on page 100.) ▼

ATTACH THE BACKER PANEL

Now you can cut the backer panel to size and secure it to the wall, as shown in the top right photo. If you want the wainscoting to fit as flush to the wall covering as possible, remove the drywall behind the panel prior to installing it.

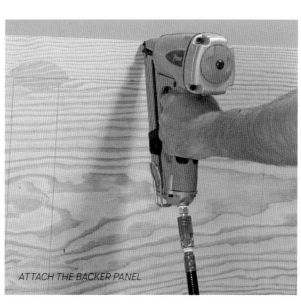

ATTACH THE BACKER PANEL

BUILD THE FRAMES

Use the dimensions you've laid out on the backer panel to cut your frame pieces to size. Then join the parts with cope-and-stick joints (page 110) or butt them together and strengthen the joints with dowels or biscuits. Assemble the frames and apply clamps as shown. Check to make sure the frame is square, and let the glue dry overnight. Repeat for the remaining frames. ▼

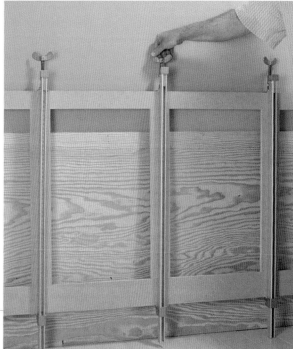

ATTACH THE FRAME

The next day, remove the clamps from the frame and position the frame on the backer panel. Secure it to the backer panel with glue and nails. Repeat for the remaining frames. ▼

ADD THE BASEBOARD

INSTALL THE CAP

The next step is to cut and install the cap or nosing on top of the wainscoting, as shown. Miter inside and outside corners. You can nail the cap to the frame and/or backer panel if you angle the fasteners. ▼

ADD THE BASEBOARD

With all the frames attached to the backer panel, work your way around the room adding baseboard, as shown in the top right photo. Cut and install either one-piece or built-up baseboard, following the directions on pages 92–97.

INSTALL THE SCOTIA

If you're planning on a scotia, now's the time to add it. Cut it to fit and nail it to the wainscoting or cap, as shown. When all the scotia is in place, go around the room, filling all nail holes. Then apply the finish of your choice. ▼

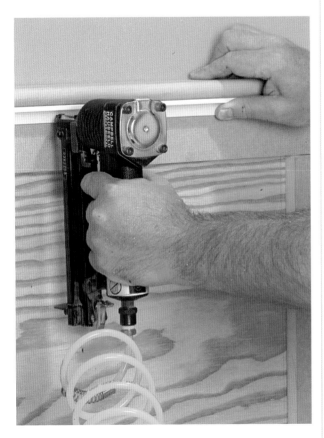

PRO-TIP: FINISHING TOUCHES

Frame-and-panel wainscoting looks just fine plain. But if you want to dress it up, consider adding either inside trim or pre-made panels. ▼

Add inside trim. A simple way to dress up frame-and-panel wainscoting is to install molding or trim around the inside perimeter of each "panel." To do this, miter trim to fit and secure it to the frames with glue and brads, as shown.

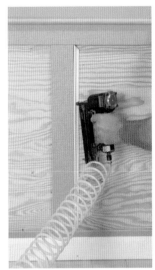

Install panels. If you'd like to add a distinctive touch to your wainscoting, consider installing pre-made raised panels onto each of the wainscoting's flat panels. The highly sculpted panels shown here are urethane foam made by Fypon (www.fypon.com). They can be secured easily with polyurethane glue and a couple of brads.

115

DRESSING UP CABINETS

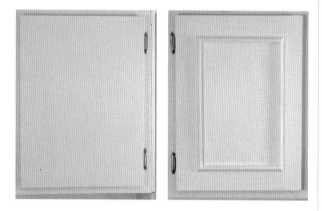

Plain doors on cabinets can be dressed up with trim in no time: Just look at the difference between the before and after photos above. Basically, what you're doing is making wall frames, as described on pages 100–101, and installing them on cabinet doors instead of walls.

Sizing frames for cabinet doors is different from wall frames, however. Instead of using the golden rectangle (page 100), it's usually best to simply mimic the shape of each door, as illustrated in the drawing below. ▼

BUILD THE FRAMES

When you've sized your frames, miter the ends as described on page 66. Then make a simple jig to assemble them. The jig is just a pair of rectangles: a form and a base. The form is cut to the exact inner dimensions of the frame and attaches to the base. This automatically squares up the frame as it's assembled. (Note that you'll have to make a set of these for the different-sized doors—start with the largest, build the frames, and then cut the jig parts down to the next size—repeat for all the frames.) Apply adhesive to the mitered ends, place the frame parts on the jig, and pin the ends together with brads, as shown. ▼

Use equal spacing to determine size of frame

Door

Attach frame to door

Cabinet

ATTACH THE FRAMES TO THE DOORS

Make sure your cabinet doors are clean before installing the frames to ensure a good bond between the adhesive and the doors. If needed, clean the doors first with TSP (tri-sodium phosphate, available wherever paint supplies are sold). Make sure to allow the doors to dry completely before proceeding. To securely fasten a frame to a door, first apply a bead of adhesive along the back edge of the frame (top photo below). Then center the frame from side to side and from top to bottom on the door and secure it with brads, as shown in the bottom photo. Repeat for the remaining frames. ▼

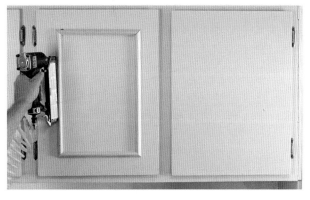

FILL THE NAIL HOLES

Once you've attached all the frames to the doors, go back and fill in all the nail holes with putty. When dry, sand the putty smooth, as shown in the top right photo.

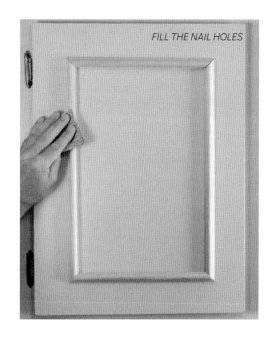

FILL THE NAIL HOLES

APPLY A FINISH

The final step is to apply a finish. You can paint the cabinet doors or apply a clear coat if you used solid wood trim to match your doors. ▼

PRO-TIP: INSTALLING CROWN ON CABINETS

Another way to dress up cabinets is to install crown molding on top. If the cabinets are near the ceiling, the crown molding can conceal the gap between the cabinet tops and the ceiling, as shown in the drawing below. You can attach crown molding made from matching wood, as shown here.

Another popular look is to use paint-grade molding (sold at most home centers), and create a striking contrast by painting the molding to match the accent colors used in the room. Crown molding can be attached to cabinets via blocking with or without an extension, as illustrated in the drawing below. ▼

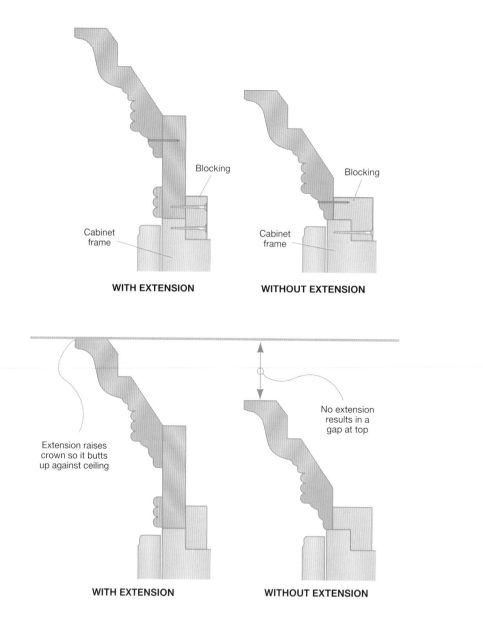

Blocking

Cabinet frame

WITH EXTENSION

Blocking

Cabinet frame

WITHOUT EXTENSION

Extension raises crown so it butts up against ceiling

No extension results in a gap at top

WITH EXTENSION

WITHOUT EXTENSION

Attach the blocking. Besides using a base molding as described on page 151, the next most secure way to attach crown molding to a cabinet top is to cut and attach support blocks. The blocks are just scraps of 2x4 that are notched to sit flush with the cabinet front, as shown. Mark along the top of the cabinets every 16" or so and at any corners, and attach a block at each mark with glue and screws. ▼

SECURE CROWN TO BLOCKING

Add the accent strip. Much of the crown molding available from cabinet manufacturers comes in two pieces: the crown molding and an accent strip that is attached to a flat on the molding. The system allows the manufacturers to offer different looks using the same base molding. Cut the accent strips to length, mitering the end or ends as needed, and secure them to the flat section of the crown molding with glue and brads. ▼

Secure crown to blocking. With the blocking in place, cut your crown to fit, as described on pages 154–155. When possible, cut miters first and then trim the molding to length, always erring on the long side. Once cut, attach the crown to the blocking, as shown in the top right photo.

INSTALLING PLATE OR PICTURE RAIL

Plate or picture rails are narrow shelves attached to walls near the ceiling to display pictures or plates. They typically consist of either shelf brackets and a shelf (as shown here), or a two-piece shelf made up of a top and a base. In either case, a groove is cut near the top front edge of the shelf or top to keep plates and pictures from accidentally sliding off, as illustrated in the drawing below. The groove also helps keep everything leaning at the same angle. Plate rail should always be mounted to wall studs with screws, since these shelves often support a lot of weight. ▼

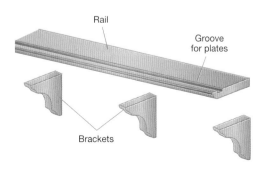

Rail

Groove for plates

Brackets

PRO-TIP: PRE-MADE BRACKETS

You can make your own brackets for plate rail as described on page 122 or use pre-made brackets like the ones shown here. Besides being convenient, pre-made brackets typically sport a built-in metal hanger recessed into their back edge that makes hanging them a snap. ▼

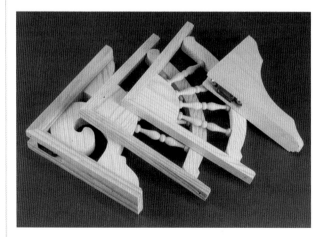

LOCATE THE RAIL AND STUDS

To install a plate/picture rail, start by measuring down the desired distance from the ceiling and make a mark. Then snap a level chalk line around the perimeter of the room as a reference for installing the rail (alternatively, use a laser level as shown here). Now use an electronic stud finder to locate the wall studs and mark their locations. ▼

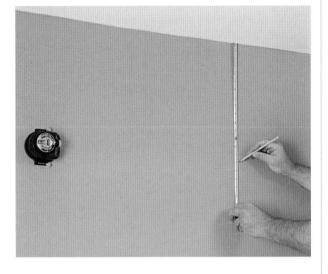

ATTACH THE BRACKETS

If you're using brackets to support the shelf, mark locations for the screws at each wall stud along the chalk or laser level line. Drill pilot holes and drive in the screws. Then position a bracket over the screw and slide it down to engage the metal hanger. ▼

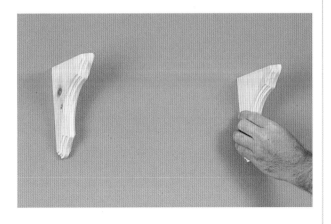

ATTACH THE RAIL

Plate rail should be made from solid wood—particleboard and MDF don't have sufficient strength to span the brackets without sagging over time. Cut your plate rail to width and then cut or rout a plate groove. Then position the plate rail on the brackets and drive screws or nails through the rail and into the brackets. When all the rail is in place, go back, fill all nail holes, and apply a finish. ▼

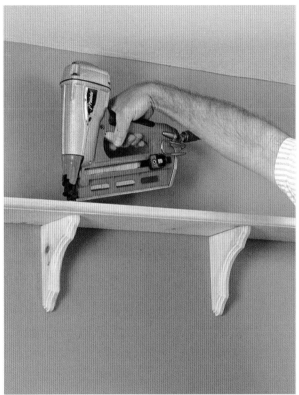

PRO-TIP:
MAKING BRACKETS

If you want to make your own brackets for plate or picture rail, you'll need to make quite a few. A router is the perfect tool to make duplicate parts—especially curved parts like the plate rail brackets shown here. Our simple technique entails making a pattern and using this to duplicate the parts. The pattern or template is made from ¼" hardboard, since its smooth surface and dense edges make it perfect for running a bearing along its edge. A special router bit called a patternmaker's bit is used with the template to duplicate parts, as described in the sidebar on the opposite page.

Trace pattern on stock.
To make duplicate brackets, start by making a pattern. You can use the pattern shown here, or create your own. Transfer the pattern to a piece of ¼" hardboard, and cut and sand it to shape. Use this to lay out the pattern on your bracket stock, as shown.

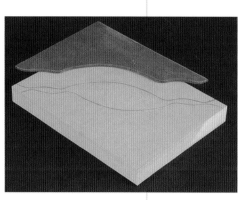

Cut to rough shape. The next step is to cut the brackets to rough shape with a coping saw, saber saw, or band saw.

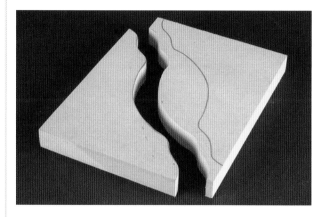

FULL SIZE

SAND TO FINAL SHAPE

There are two options for creating the final bracket shape from the rough-sawn blank: You can sand it to shape (as shown here) or, to make identical brackets, you can pattern-rout them as described in the sidebar at right. ▼

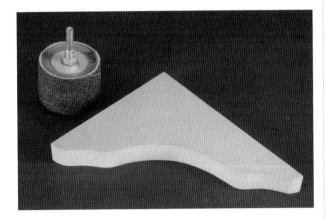

ROUT THE EDGES

All that's left is to rout a roundover or decorative profile on the front edges of each bracket, as shown below. ▼

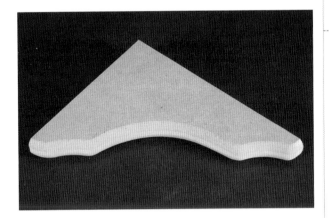

PRO-TIP: PATTERN-ROUTING

The key to pattern-routing is a special router bit called a patternmaker's bit. It's basically a straight bit with a bearing mounted on top; the bearing is the same diameter as the bit. By attaching a template to the top of a workpiece, you can guide the bearing along the edge of the template; the straight bit will trim the workpiece to the identical shape of the template, as illustrated in the drawing below. ▼

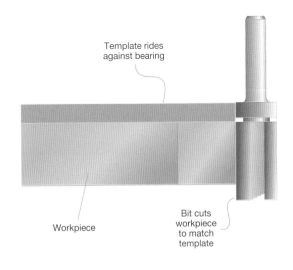

Template rides against bearing

Workpiece

Bit cuts workpiece to match template

123

ADDING PILASTERS

A pilaster is a shallow column that's attached to a wall or cabinet to create the illusion that it's providing support. A pilaster can run from floor to ceiling or partially up a wall. Some versions come in two pieces: a pilaster and a base—the pilaster can be cut to length and then the base is added, as illustrated in the drawing at right. With the advent of easy-to-use urethane foam versions, pilasters can readily be incorporated into any home's interior or exterior. The pilaster we installed here is manufactured by Fypon (www.fypon.com). It's important to note that decorative urethane trim like this does not provide any structural support.

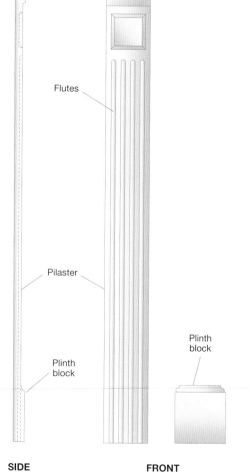

Flutes

Pilaster

Plinth block

Plinth block

SIDE VIEW

FRONT VIEW

124

LOCATE THE STUDS

To install a pilaster, begin by locating the wall studs. Use an electronic stud finder to locate these, and then mark each location with a pencil. ▼

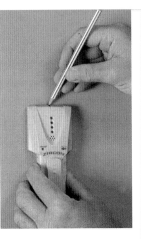

CUT THE PILASTER TO LENGTH

If your pilaster does not have a separate base (as shown here), cut your pilaster about $1/16$" less than the floor-to-ceiling measurement. This will provide enough clearance to position the pilaster. For a two-piece pilaster, you can cut it $1/4$" or so less than the measurement. This will make it easier to install, and the separate base piece will cover up any gap. ▼

MEASURE THE WALL

If your pilaster is designed to run from floor to ceiling, the next step is to measure this height. ▼

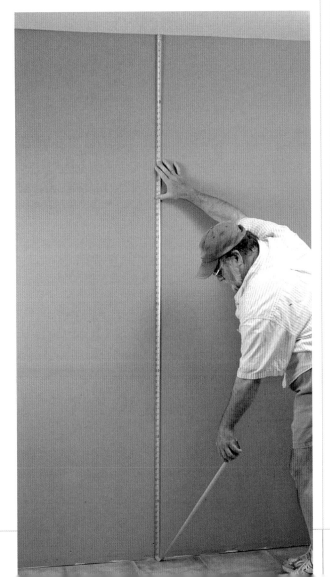

FLOOR AND WALL TRIM

APPLY AN ADHESIVE

For a urethane foam pilaster, apply a bead of quality urethane adhesive to its back. For a wood pilaster,
apply a bead of construction adhesive. ▼

INSTALL
THE PILASTER

ATTACH THE BASE

All that's left is to slip the base (if applicable) over the
pilaster and secure it with nails. Fill any nail holes with
putty and apply the finish of your choice. Repeat for
any remaining pilasters. ▼

INSTALL THE PILASTER

Position the pilaster in the desired location so it
butts up against the ceiling, and check it for plumb
with a level. Then secure it to the wall as shown in
the top right photo. If possible, drive fasteners into
a wall stud.

PRO-TIP: MAKING PILASTERS

You can make your own pilasters if you have a router and a table saw. Most pilasters sport a series of flutes—shallow half-moon-shaped grooves—that run the length of the pilaster and are spaced evenly across its width, as illustrated in the drawing below. Although flutes are usually cut with a core-box bit, they can also be cut with a V-groove bit. The only trick to cutting these is spacing the grooves evenly. The easiest way to do this is with spacers, as described below. ▼

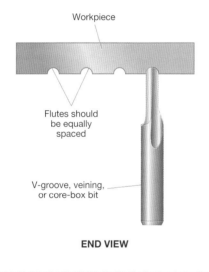

Workpiece

Flutes should be equally spaced

V-groove, veining, or core-box bit

END VIEW

Fluting setup. Precisely spaced flutes are easy when you use spacers and a fence on the router table, as illustrated in the top right drawing. The advantage of using spacers is that you don't have to reposition the fence after every flute is routed. Instead, you simply add or remove spacers. This guarantees accurate spacing, as long as the spacers are all cut to the same width. To set up for routing the flutes, start by placing all of the spacers between the bit and router table fence. Then butt them all firmly up against the fence. As you rout flutes, you'll remove the spacers one at a time.

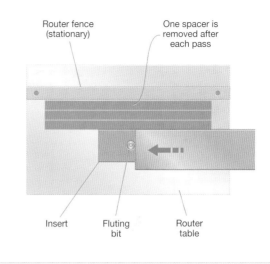

Router fence (stationary)

One spacer is removed after each pass

Insert

Fluting bit

Router table

Routing flutes. To rout flutes, adjust a core-box bit or V-groove bit for the desired cut. Then position the fence and spacers as described above. Turn on the router and rout the first flute. Then remove a spacer and rout the second. Continue like this until all the flutes are routed, as shown. ▼

CHAPTER 5:

WINDOW AND DOOR TRIM

There's a visibility problem with trim on windows and doors, both inside and out: We're so used to it that we don't even see it. But when done right, window and door trim can be a decorative element. It can follow the architecture of an exterior, mimic a style in an interior, or make a bold, contrasting statement. In this chapter, we'll show you how to change the look of a room or an exterior simply by installing window and door trim.

CASING BASICS

The variety of window and door trim, or casing, as it's often called, is almost as varied as styles of windows and doors available—everything from plain, simple molding to high-profile fancy trim.

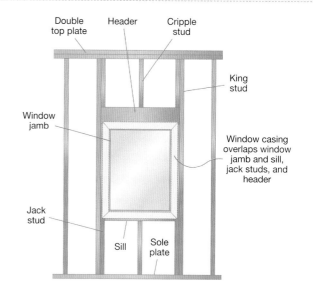

NAILING FOUNDATION

One reason that window and door casing is fairly easy to install is that there's an excellent nailing foundation, as illustrated in the drawing at right. Unlike other trim (such as wainscoting, chair rail, and crown molding) where you can only attach the trim every 16" or so due to stud spacing, window and door framing offers a continuous nailing foundation. ▶

CASING NAILING PATTERN

It's important to understand that casing covers the gaps between the rough framing and the window or door jamb, as illustrated in the bottom drawings. You'll usually use two different-sized nails to attach casing: a 2" or 2½" casing nail to secure the casing to the framing, and a 1½" finish nail to attach the casing to the window or door jamb. If you try to attach the casing by driving fasteners in the middle of the casing, you'll often hit the gap, and the fastener won't hold the casing in place. ▼

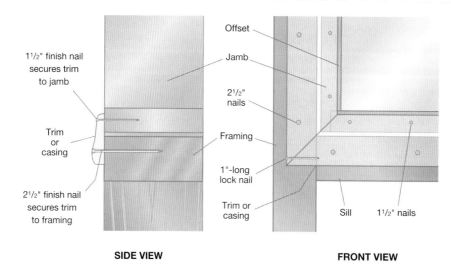

SIDE VIEW   **FRONT VIEW**

Casing is typically offset from a window or door jamb, as illustrated in the drawing below. This offset—often called a reveal—does a couple of things. It helps conceal any variations in the trim and/or the jamb, and it provides a shadow line for visual interest. There are two ways to mark a reveal: with a combination square and with a finger gauge (see below). ▼

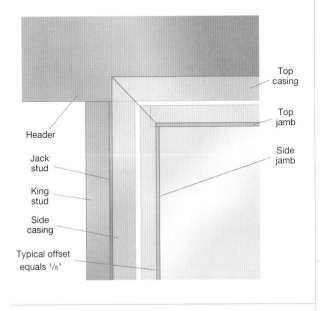

Top casing

Top jamb

Side jamb

Header

Jack stud

King stud

Side casing

Typical offset equals 1/8"

COMBINATION SQUARE

COMBINATION SQUARE

The easiest way to mark a reveal is to set the blade of a combination square so that it protrudes 1/8". Then place a pencil against the blade of the square, press the head of the square against the jamb, and run these around the perimeter to mark the reveal, as shown in the top photo.

FINGER GAUGE

If you don't have a combination square, you can use your fingers as a simple gauge. Hold the pencil in your hand, as shown in the bottom photo, and use your fingers as a stop to position the pencil so that it extends out the desired amount. Then slide your hand down the jamb to mark the reveal.

FINGER GAUGE

INSTALLING WINDOW TRIM

Installing window trim, or "trimming out" a window, can be done several ways. The two most common are picture-framing and stool-and-apron, as illustrated in the drawing at right. ▶

PICTURE-FRAMING

The most common style for attaching trim in modern homes is picture-frame trim, so called because that's what it looks like: the frame of a picture. Miter-cut trim can be a challenge to install, as the 45-degree miter cuts must be accurate for the frame parts to come together with no gaps. The advantage to this style trim is that no end grain is exposed—the wood grain appears to run continuously around the perimeter of the window.

STOOL-AND-APRON

Common in older homes, stool-and-apron casing consists of a top, two side pieces, a stool, and an apron. The main advantage to this system is that since all the cuts are square and the pieces are butted together, it's easy to install. The disadvantage is that this leaves wood grain exposed on the trim pieces that show on the sides of the window. One way to get around this is with a mitered return; see the sidebar at right.

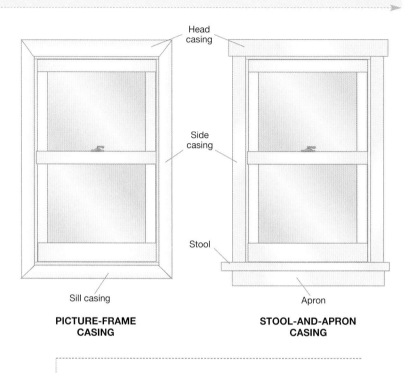

Head casing

Side casing

Stool

Sill casing

Apron

PICTURE-FRAME CASING

STOOL-AND-APRON CASING

PRO-TIP: MITERED RETURNS

Instead of cutting a butt joint on a top, stool, or bottom casing in stool-and-apron trim, leaving the end grain exposed, pros make a mitered return. For more on making a mitered return, see page 94. ▼

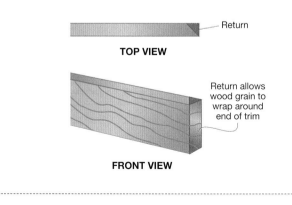

Return

TOP VIEW

Return allows wood grain to wrap around end of trim

FRONT VIEW

MARK THE REVEALS

To trim a window, start by marking the reveals, as shown in the top photo and described on page 131.

INSTALL THE CASING

The sequence for attaching trim depends on the type of trim you're using, picture-frame or stool-and-apron, as illustrated in the bottom drawing. Position a piece of casing so its inside edge is flush with the marked reveal. Then make marks at the top and bottom of the trim where the horizontal and vertical reveal lines meet. Cut the trim at the marks and install the trim, as shown in the bottom photo. Once this piece is installed, move on to the next piece. Note: If your window isn't square, start with the casing long and sneak up on the cut, adjusting the saw angle as necessary.

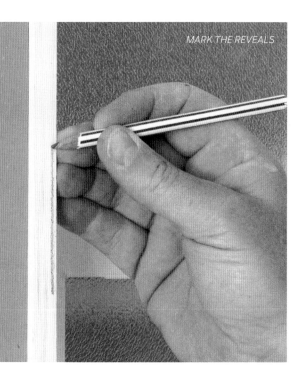

MARK THE REVEALS

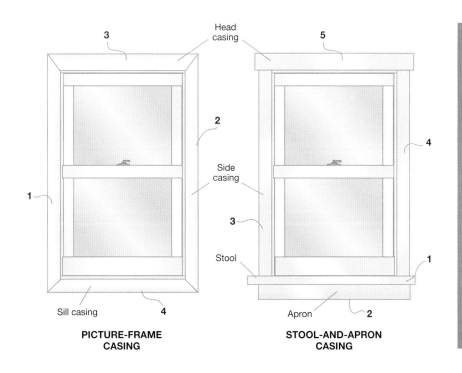

PICTURE-FRAME CASING

3

Head casing

2

1

Sill casing   4

STOOL-AND-APRON CASING

5

Side casing

4

3

Stool

Apron   2

1

INSTALL THE CASING

PRO-TIP: EXTERIOR WINDOW/DOOR TRIM

Interior window and door casing conceals the gaps between the wall coverings and the window and door jambs. Exterior trim does the same thing, and also helps create a weather-tight seal, as illustrated in the drawing below. First, it holds in or conceals insulation that's installed in the gaps between the window or door jamb and the framing. Then once the trim is in place, the edges are sealed with silicone caulk to further help prevent drafts and keep out moisture. ▼

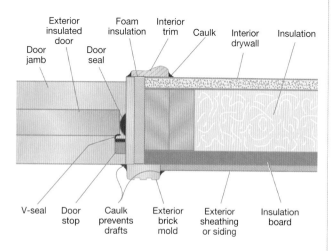

Exterior insulated door — Foam insulation — Interior trim — Caulk — Interior drywall — Insulation

Door jamb — Door seal

V-seal — Door stop — Caulk prevents drafts — Exterior brick mold — Exterior sheathing or siding — Insulation board

Window mounting options. You'll often find it tougher to drive in casing nails on new-construction windows because they have to pass through the nailing fins. Windows are mounted through the jambs or via nailing fins, as illustrated in the drawing below. Most new-construction windows come with perimeter flanges that are nailed to the exterior sheathing, as shown. Traditional windows are attached by driving screws or nails through the jambs and shims into the rough opening framing. ▼

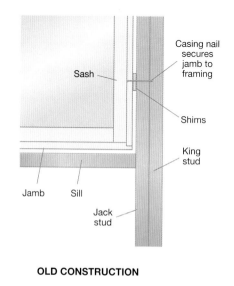

Casing nail secures jamb to framing

Sash

Shims

King stud

Jamb — Sill

Jack stud

OLD CONSTRUCTION

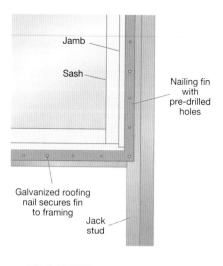

Jamb

Sash

Nailing fin with pre-drilled holes

Galvanized roofing nail secures fin to framing

Jack stud

NEW CONSTRUCTION

Add insulation. On exterior windows, the gaps between the framing and window jambs should be filled to prevent drafts, usually with foam insulation or fiberglass. When using insulating foam (as shown in the photo), remember to use the minimal-expanding type designed for windows and doors. If you use the wrong type or too much foam, the expansion can bow the jamb, resulting in a window that sticks. Fiberglass insulation also does a good job of preventing drafts. Just remember that insulation works by trapping air; if you pack it in too tight, the insulation can't do its job. ▼

Install the casing. There are two ways to install exterior trim to a window: before or after the exterior covering is in place. It's easiest to install casing before the wall covering, since the covering is then simply butted up against the casing and sealed with caulk (see below). For most remodel work, the casing is installed after the exterior covering is in place, as shown in the top right photo. This casing is installed much like exterior casing (pages 132–133), except that you'll use galvanized fasteners and you'll need to caulk carefully, as described below.

INSTALL THE CASING

Seal the window. Once your exterior trim is in place, take the time to apply 100% silicone caulk around the perimeter of the trim, as shown, to further seal against drafts and to keep out the elements. ▼

135

INSTALLING DOOR TRIM

By its very nature, door trim is simpler to install than window trim because it has fewer parts: just two sides and a top. This also means that you only have to cut two miter joints where the top pieces meet. The bottom ends of the side casing are cut square. ▼

MARK THE REVEALS

To install picture-frame door casing, start by marking the reveal on each jamb, as shown here and described on page 131. ▶

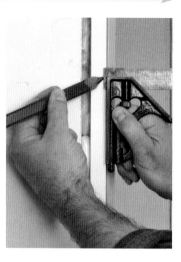

CUT AND INSTALL THE TRIM

Most trim carpenters will start with the vertical trim. Place a piece of trim in position so its inside edge is flush with the marked reveal. Then make marks at the top and bottom (if applicable) of the trim where the horizontal and vertical reveal lines meet. Once these are cut and installed, move on to the top trim piece. Measure for this by running a tape measure between the installed side pieces, and cut a top piece to fit. Note: If your door isn't square, start long and sneak up on the cut, adjusting the saw angle as necessary. Attach the casing to the framing that surrounds the window, as shown here and described on page 130. ▶

LOCK-NAIL THE MITER JOINTS

Because miter joints tend to open and close as the humidity changes, it's a good idea to "lock-nail" the miters, as shown in the top photo. Simply drive a nail in through the top or side of the miter joint into the adjacent piece to lock the pieces together. Since you'll be nailing close to the edge, it's best to drill a small hole first to prevent the nail from splitting the trim.

FILL ALL NAIL HOLES

With the casing in place, go back and fill in the nail holes with putty. For exterior trim, make sure to use putty that's formulated for exterior use and one that will dry hard. Many types of putty never harden and should not be used outdoors. Press the putty into a hole with your finger, as shown, or with a putty knife. Be sure to overfill the hole slightly, as most putty will shrink as it dries. When dry, the putty will likely stand proud of the trim surface and need to be sanded flush. Use an open-coat sandpaper for this—it has less tendency to clog than most sandpaper.

SEAL WITH CAULK

Finally, use paintable latex caulk to fill in any gaps between the trim and the interior wall (use 100% silicone caulk to seal the exterior trim). Likewise, caulk any gaps between the trim and the jamb. Smooth the caulk with a wet finger, and when dry, paint the trim with a quality oil-based trim paint. ▼

PRO-TIP: EXTERIOR DOOR TRIM

Although most exterior door trim is brick mold installed like any picture-frame casing, it can be more elaborate, as illustrated in the drawing below. This type of trim (described in detail on pages 173–175) typically consists of side pilasters that terminate in plinth blocks, and an elaborate header complete with keystone. ▼

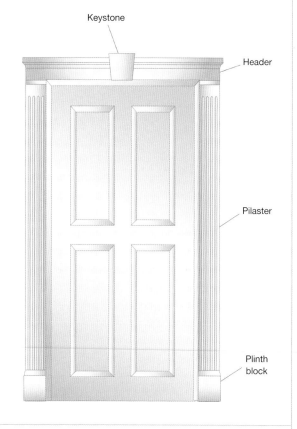

Keystone

Header

Pilaster

Plinth block

Brick mold or brick molding is much thicker than interior door casing. That's because, on homes with brick exteriors, it was originally used to provide an edge for bricks to butt up against. Although it's used for all types of exteriors now, the name has stuck—along with the unnecessary thickness. There are two ways to install brick mold: before or after the door is installed, as described here and on the opposite page.

Brick mold prior to installation. Most door installers prefer to attach brick mold to the door jamb prior to installing a door, as shown. That's because the brick mold will serve as a lip to prevent the door from passing all the way through the opening as it's being set in place. As with any door casing, mark the reveals first (page 131), cut the casing to fit, and secure it to the door jamb so it aligns with the reveals. ▼

Brick mold after installation. For remodel work where you are replacing or upgrading exterior door casing, you'll be attaching the casing with the door in place, as shown. The only problem that may arise here is that the new casing is not the same width as the original. If it's not, you'll have to either shim it out or cut away the exterior wall covering to get it to fit. ▼

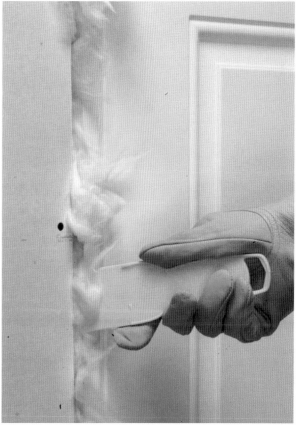

Insulate with fiberglass. Fiberglass insulation also does an admirable job of preventing drafts. Again, the thing to remember here is that insulation works by trapping air. If you pack the fiberglass in too tightly, the insulation can't trap air, and it won't provide any insulation properties. ▼

139

Insulate with foam. On exterior doors, before the door casing is installed, the gaps between the framing and the door jambs should be filled to prevent drafts. When using insulating foam (as shown in the top right photo), remember to use the minimal expanding type designed for windows and doors. If you use the wrong type or too much foam, the expansion can bow the jamb, resulting in a door that sticks.

TRIMMING A DOOR OPENING

In many homes with pass-through door openings, the walls are often left plain. Typically the opening is finished with drywall only, with no attempt at trimming the opening to match the surrounding room interiors. In years past, trimming an opening required considerable woodworking skill. But with the pre-made parts (like plinth blocks, fluted trim, and rosettes) that are now commonly available at most home centers and lumberyards, trimming an opening is easy. ▼

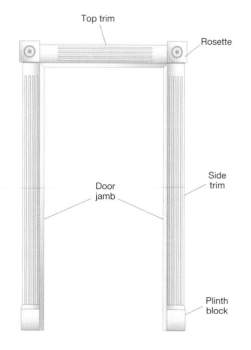

Top trim

Rosette

Door jamb

Side trim

Plinth block

INSTALL THE JAMB

In addition to providing a finished look for the wall opening, a jamb also provides a nailing surface for the wall trim you'll add later. Pre-made jambs are available wherever doors and trim are sold. Most jamb sets accommodate a range of door widths and heights. Odds are that you'll need to trim the parts to length to fit your door opening. Position the top jamb between the side jambs, and fasten them together with nails. Then slide the jamb into the opening and secure it to the framing members, as shown. ▶

INSTALL THE PLINTH BLOCKS

With the jamb installed, the next step is to attach plinth blocks at the base of the opening. These can either be installed flush with the jamb or offset $1/8"$ or so—it's really a matter of personal preference. ▼

ATTACH THE ROSETTES

The rosettes are installed next to the top corners of the jamb. Make sure that these go in level and plumb with the plinth blocks. A plumb bob is a great way to check alignment. Alternatively, if you offset the plinth block, you can mark and use the same offset or reveal on top of the jamb for the rosettes. ▼

INSTALL THE SIDE TRIM

Once the plinth blocks and rosettes are in place, the next step is to install the side trim. Measure from plinth block to rosette, and cut pieces of trim to fit. Install these so they are centered on the width of the plinth blocks and rosettes. Fasten the trim to the jamb with nails, as shown in the top right photo.

INSTALL THE SIDE TRIM

ATTACH THE TOP TRIM

All that's left is to cut and install the top trim piece. Measure from rosette to rosette, and cut a piece to fit. Center it on the rosettes and attach it with nails, as shown. Repeat this process for the other side of the opening. Then fill all nail holes and apply the finish of your choice. ▼

CHAPTER 6:
CEILING TRIM

The last feature we tend to notice in a room is the ceiling...because usually there's nothing much to notice. Painted white or off-white, flat or spray-textured, ceilings tend to be ignored. But with all that real estate overhead, a ceiling can have an enormous effect on the look of a room. In this chapter we'll show you how you can add some snap to your ceiling with simple molding, ceiling medallions, crown molding, and even ceiling paneling.

BASIC INSTALLATION SEQUENCE

All ceiling trim is installed using the same sequence. The typical sequence for installing perimeter trim is illustrated in the drawing below. Start by measuring the longest, most visible wall. Cut a piece of trim to fit this and butt it into the corners as shown. To achieve the best possible appearance, all inside corner joints should be coped as described on pages 70–71. Outside corner joints are mitered, as described on pages 62–69.

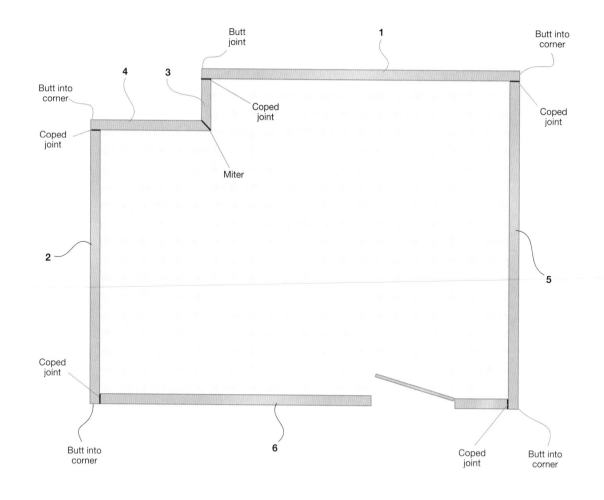

SIMPLE CEILING MOLDING

Simple molding—even when painted the same color as the ceiling—can add a lot of visual interest to an otherwise ordinary room. Simple ceiling trim varies from cove or quarter-round to more complex shapes. Depending on the type, it may be installed flat against the wall or angled, as shown here. ▼

LOCATE AND MARK WALL STUDS

ATTACH WITH FASTENERS

Follow the installation sequence described on page 144 to measure and cut trim to length. Butt the first piece of trim into one corner on the longest wall, and drive fasteners through the trim and into the wall studs and ceiling joists if possible. ▼

LOCATE AND MARK WALL STUDS

To install ceiling molding, start by using an electronic stud finder to locate the wall studs. Then mark these with a pencil, as shown in the top right photo. If you're installing angled trim, it's also a good idea to locate and mark the ceiling joists as well so that you can secure the trim to them when possible.

SCARF AS NEEDED

As you work your way around the room, chances are that you'll need to fasten strips of molding together to cross long expanses. When this occurs, miter the ends of the molding in opposite directions to create a nearly invisible "scarf" joint. At inside corners, cope the joints as described on pages 70–71, and miter any outside corners. ▼

FILL THE NAIL HOLES

Once all the trim is in place, go back around the room and fill all the nail holes with putty, as shown in the top right photo. Fill these just slightly proud of the surface. When the putty is dry, go back and sand the filled holes flush with the molding surface.

CONCEAL ANY GAPS

Finally, fill any gaps between the trim and the ceiling or wall with a good-quality paintable latex caulk. Wipe away any excess immediately, and when dry, mask and paint the crown molding if desired. ▼

PRO-TIP: FOAM TRIM

Urethane foam is easy to cut, very lightweight, and quick to install. The trim shown here is manufactured by Fypon (www.fypon.com) and comes in a dizzying array of profiles. Installing foam trim is similar to installing wood trim, with the following exceptions. First, foam trim should be used for decorative purposes only; it does not provide any structural support. Second, it's important to check your local building codes to make sure the trim meets your local specifications. Third, you can paint this trim any color you want. It takes paint well, and some manufacturers make embossed trim that looks like wood, so you can stain it to match other wood in the room.

Always use an adhesive. Foam trim should never be installed with just fasteners. It's designed to be installed with both fasteners and a bead of high-quality, urethane-based adhesive. You don't need a lot here: About a $1/8$" bead on both mating surfaces will work fine. You'll find that this combination of adhesive and fasteners, along with the foam molding's built-in flexibility, will make installing molding on even walls a cinch.

Cut long. To install urethane foam trim, measure for the piece, add $1/8$", and cut the trim to this length. Yes, the manufacturer actually suggests cutting the trim a bit long so you end up with a nice, tight friction-fit. Unlike unforgiving wood trim, foam trim will actually compress a bit as needed.

Secure the trim. To secure the trim, start by applying a bead of urethane adhesive to the contact points or "flats" of the trim, as shown in the top photo. Then apply a bead to the ends of the trim, as shown in the middle photo. Now position the trim and secure it by driving nails through the trim and into the wall studs, as shown in the bottom photo. Repeat for the remaining walls.

CEILING MEDALLIONS

A ceiling medallion can transform a room from ordinary to extraordinary. Medallions are easy to install and come in a huge array of styles and sizes. The medallion shown here is made by Fypon (www.fypon.com) and is shaped from urethane foam, so it's lightweight and quick to install. Medallions can also tackle a common problem when replacing an overhead light fixture: If the base plate of the new fixture is smaller than the old one, it can leave an unsightly portion of the ceiling exposed. You could patch and paint this area, but a medallion will cover up the problem and also add a distinctive touch to both the ceiling and the new light fixture.

PAINT IF DESIRED

Urethane foam takes paint readily, so if you'd like to add some color, paint the medallion now, when it's easiest, before mounting it to the ceiling. ▼

APPLY AN ADHESIVE

Because it's so light and offers such a large gluing surface, a urethane medallion can be secured to the ceiling just by applying a bead of high-quality urethane adhesive to its back. ▼

SECURE WITH FIXTURE

To install the medallion, simply press it in place over the electrical box, as shown in the inset photo. Apply a couple of strips of tape to keep it in place until the adhesive sets up. Then install the fixture, as shown below right. ▼

INSTALLING CROWN MOLDING

Crown molding can dress up any room, adding a graceful touch with its classic profiles. Crown molding comes in many different profiles and sizes, as illustrated in the drawing below.

DESIGN GUIDELINES

As a general rule, you should choose crown molding that's around 3" to 4" wide for a standard 8' ceiling. Anything wider will appear out of scale. Also note that crown molding comes in varying lengths up to 16' and is sold by the foot. If you can't find molding that's long enough to span a room from corner to corner, you'll have to join the pieces together with a scarf joint.

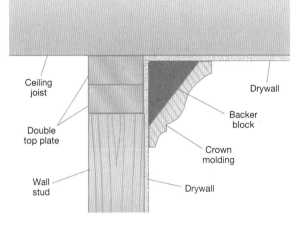

Ceiling joist

Double top plate

Wall stud

Drywall

Backer block

Crown molding

Drywall

SPRING ANGLE

The angle at which crown molding fits up against the wall and ceiling can also be specified, as described on page 155. The two most common variations are a 52/38-degree spring angle and a 45/45-degree spring angle.

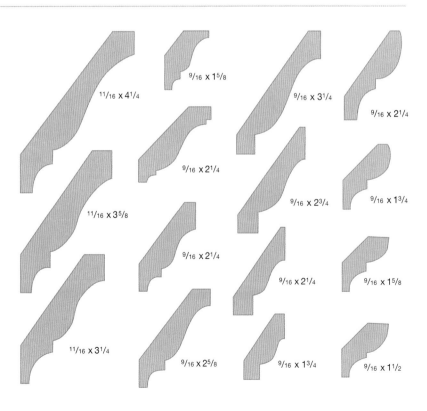

$^{11}/_{16}$ x $4^{1}/_{4}$

$^{9}/_{16}$ x $1^{5}/_{8}$

$^{9}/_{16}$ x $3^{1}/_{4}$

$^{9}/_{16}$ x $2^{1}/_{4}$

$^{9}/_{16}$ x $2^{1}/_{4}$

$^{9}/_{16}$ x $2^{3}/_{4}$

$^{9}/_{16}$ x $1^{3}/_{4}$

$^{11}/_{16}$ x $3^{5}/_{8}$

$^{9}/_{16}$ x $2^{1}/_{4}$

$^{9}/_{16}$ x $2^{1}/_{4}$

$^{9}/_{16}$ x $1^{5}/_{8}$

$^{11}/_{16}$ x $3^{1}/_{4}$

$^{9}/_{16}$ x $2^{5}/_{8}$

$^{9}/_{16}$ x $1^{3}/_{4}$

$^{9}/_{16}$ x $1^{1}/_{2}$

PRO-TIP: CROWN FOUNDATIONS

Crown molding fits against a wall at an angle and has only two small flat sections that make contact. This makes it a challenge to attach, especially on walls and ceilings, since you need to hit a stud or joist for the nail to hold. To get around this, many trim carpenters use one of three methods to create a solid nailing base: angled nailing blocks, plywood blocking, or a flat base molding (see below). To install any of these, you'll first need to locate the wall studs with an electronic stud finder, and then mark these locations with a pencil.

Nailing blocks. The simplest way to provide better support to crown molding is to cut a set of angled nailing blocks and attach these to the wall studs, as shown in the bottom left photo. To make these, just measure the flat portion on the back of the crown molding. Then miter-cut scraps of 2x4 to create an identical flat on the long side of the blocks. Attach these at the marked wall studs: You'll find that it's much easier to install and attach the molding to the nailing blocks than directly to the wall studs.

Beveled strips. To create even better support for crown molding, consider ripping beveled strips of plywood to fit behind the molding. In most cases, two beveled pieces of plywood stacked on top of each other will provide a very stable nailing base. Attach the plywood blocking to the wall studs at each marked location, as shown in the bottom middle photo. Now you can attach the crown molding to the plywood blocking anywhere along its length. This is especially useful when you need to join together short pieces of molding with scarf joints—there will always be something to nail into with plywood strips.

Flat base. Another way to provide continuous nailing support is to first attach a flat molding to the wall, as shown in the bottom right photo. The flat molding is easy to attach to the wall at the stud locations and provides a continuous nailing surface for the crown molding. In addition to providing a nailing surface, these moldings can also provide a more complex and pleasing profile (especially if you shape the bottom edge with a router before installing the molding).

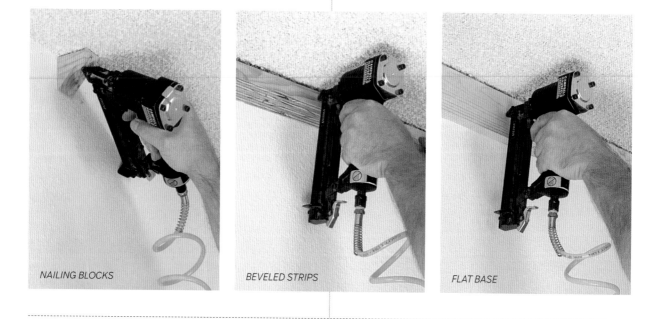

NAILING BLOCKS

BEVELED STRIPS

FLAT BASE

BUTT THE FIRST PIECE

Follow the installation sequence described on page 144 to measure and cut trim to length. Butt the first piece of trim into one corner on the longest wall, and drive fasteners through the trim and into the wall studs, as shown. ▼

COPE THE INSIDE CORNERS

With the first strip of crown in place, expose the miter on the end of the next piece and use a coping saw to remove the waste, as described on pages 70–71. Test the fit of the coped end against the molding that's butted into the corner (as shown in the top right photo), and fine-tune the coped end as needed for a good fit (see pages 80–81).

COPE THE INSIDE CORNERS

SECURE THE MOLDING

Once you've obtained a good fit between the strips of molding, you can secure the molding as shown below. Start by securing the molding to the studs, mitered angle blocks, plywood blocking, or flat base molding. If you're securing the molding directly to the studs, try to wiggle the molding after each nail is driven in to make sure you hit the stud. Drive in another nail if it's loose. ▼

PRO-TIP: CUTTING CROWN MOLDING

Although attaching crown molding is fairly straightforward, cutting the complex molding to wrap around a room—especially around corners that aren't 90-degree—can be tricky. Fortunately, with the aid of a power miter saw, this can be done without too much head scratching. There are three basic methods for cutting crown molding: with the molding flat on the table, with the molding angled against the fence, and with the aid of a jig.

But even with a tuned miter saw and a lot of patience, cutting crown can still be tricky. This is because most walls and ceiling aren't 90 degrees to each other, so you'll have to tweak the angles to make the crown fit well, as described on page 155. It's also a good idea to practice your first few cuts on scrap until you get the hang of it. Additionally, it's best to cut each piece a bit "fat" so you have some room to tweak the angles as needed.

Upside down and backwards. We prefer cutting crown angled against the fence, as this method requires the fewest mental gymnastics. Still, the standard phrase for cutting crown molding is "upside down and backwards." That is, when you position the workpiece on the miter saw, you want to turn the molding upside down and flip it end-for-end. The simplest way to handle this concept is to label the table and fence on your miter saw, as illustrated in the drawing below. ▼

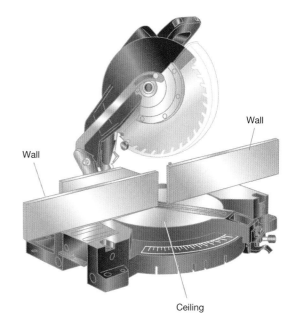

Wall

Wall

Ceiling

Using crown stops. If your saw has a crown stop (like the Bosch saw shown below), adjust the stop to hold the molding at the correct angle. To do this, just slide the stop in or out until the top flat rests flush with the fence and the bottom flat is flush with the saw table. The stop will hold the molding at the perfect angle, as shown in the bottom photo. If you don't have crown stops on your saw, you'll need to either clamp a strip across the saw table to hold the crown in place, or use a crown molding jig like the one described at right. ▼

Using a crown molding jig. If you want to cut crown molding with the molding angled up against your fence and your saw doesn't have built-in stops to hold the molding in place, consider using a crown molding jig. Crown molding jigs do one thing and they do it well—hold your workpiece at the correct angle for the cut. There are a number of crown molding jigs on the market. The crown jig shown below made by Woodhaven (www.wood-haven.com) attaches to their fence system. It holds moldings at the correct angles so all you have to do is make a miter cut. No need to fuss with bevel angles—and it works on either side of the blade. ▼

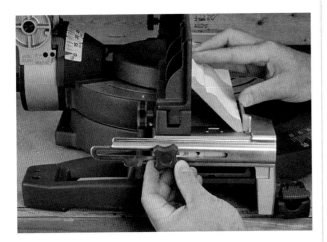

ATTACH TO CEILING JOISTS WHEN POSSIBLE

If the ceiling joists in the wall you're working on run perpendicular to the wall, it's also a good idea to secure the crown molding to these joists, as shown below. This will help hold the molding in place and prevent it from sagging over time—especially on long expanses of trim. ▼

PIN THE MITERS

If you have to wrap around an outside corner (as shown in the top right photo), it's a good idea to pin together the corners by driving a nail or two into the top edges of the molding, as shown. This will do two things: It will help pull the miter joint closed and will also create a stronger installation.

PIN THE MITERS

BURNISHING TIP

If you notice small gaps when you miter together an outside corner, you can close the gaps with an old carpenter's trick called "burnishing." All you have to do is press the shank of a screwdriver firmly over the miter joint, as shown below. This will crush the wood fibers and fill in the gap. ▼

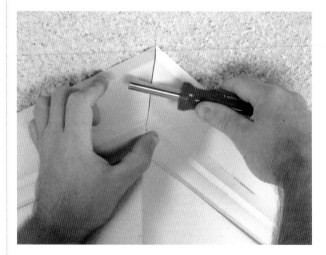

PRO-TIP: CUTTING CROWN FLAT

You usually get the most accurate cut on a miter saw with the workpiece flat on the saw table—this provides the most stable foundation. This is also true for cutting crown molding. The angle at which crown molding fits up against the wall and ceiling is called its spring angle. The two most common variations are 52/38-degree spring angle and 45/45-degree spring angle. If you don't know your spring angle, see below.

Determine the spring angle. It's important that you know the spring angle of your molding, since this will determine how you set your table angle and bevel tilt. If you don't know the spring angle of your molding, use an adjustable angle guide, as shown in the top photo, to measure it. Make sure to butt the guide up against the bottom flat of the molding. The molding we're cutting here is 52/38-degree spring molding. For 45/45-degree molding, see the chart below for the correct miter and bevel angle settings.

Adjust the saw and cut. To get ready to cut crown flat on the table, begin by setting the bevel angle to the desired angle (in our case here, 33.9 degrees). Then angle the table to the desired angle (in our case 31.6 degrees). With the saw angles set and the crown molding firmly clamped

DETERMINE THE SPRING ANGLE

to the saw table, go ahead and make the cut as shown below. You'll want to use a slow feed rate here to prevent the blade from deflecting during the cut. ▼

FLAT CROWN SETTINGS

	38-degree spring		45-degree spring	
Inside corner	Miter angle	Bevel angle	Miter angle	Bevel angle
Use the left end of cut	31.6 degrees right	33.9 degrees left	35.3 degrees right	30.0 degrees left
Use the right end of cut	31.6 degrees left	33.9 degrees right	35.3 degrees left	30.0 degrees right
Outside corner	Miter angle	Bevel angle	Miter angle	Bevel angle
Use the left end of cut	31.6 degrees left	33.9 degrees right	35.3 degrees left	30.0 degrees right
Use the right end of cut	31.6 degrees right	33.9 degrees left	35.3 degrees right	30.0 degrees left

PRO-TIP: TWO-PIECE CROWN

Much of the crown molding available from cabinet and trim manufacturers comes in two pieces: the crown molding body and an accent strip that is attached to a flat on the molding, as illustrated in the drawing below. The system allows the manufacturers to offer different looks using the same crown molding. It also lets you change the look of the crown molding later by simply removing the old accent strips and installing new strips for a bold, new look. ▼

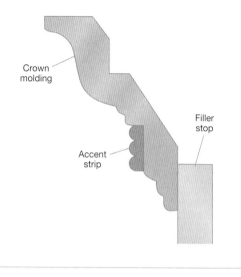

Crown molding

Accent strip

Filler stop

Install the crown body. To install two-piece crown molding, use the same procedure as for single-piece crown described on pages 149–151. Secure the crown molding body with fasteners, as shown in the top right photo.

INSTALL THE CROWN BODY

Attach the accent strip. With all the crown molding body in place, go ahead and cut the accent strips to length, mitering the end or ends as needed. Secure the strips to the flat section of the crown molding with glue and brads. ▼

PRO-TIP: NO-MITER TRIM

If you can cut a piece of trim so that it's square on the end, you can install ceiling trim. The no-miter urethane foam trim system shown here makes installing ceiling trim a snap, with its pre-made corner and divider blocks. It's manufactured by Fypon (www.fypon.com). In addition to lengths of trim, their system includes three different blocks: outside corners, inside corners, and divider blocks, as illustrated in the drawing below. These are available in numerous styles, with matching trim. ▼

CROWN MOLDING

OUTSIDE CORNER

INSIDE CORNER

DIVIDER BLOCK

Secure the blocks in the corners. What's really nice about using the pre-made blocks is that you simply glue them into the corners using a high-quality urethane adhesive, as shown. Also, since the block profile is slightly larger than the crown molding profile, it creates an additional shadow line that looks especially good. ▼

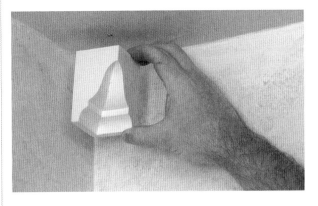

Butt joint trim. After you've installed all the corner blocks, measure the distance between them, add ⅛", and cut a piece of crown to this length. Yes, the manufacturer actually suggests cutting the molding a bit long so that you end up with a nice, tight friction-fit. Unlike unforgiving wood molding, the foam molding will actually compress a bit as needed. To install the molding, start by applying a bead of urethane adhesive to the end of the molding. Then position the molding and secure it by driving nails through the molding and into the wall studs, as shown. Repeat for the remaining walls. ▼

CEILING PANELING

A bad-looking ceiling can be transformed with good-looking paneling, especially if you opt for today's wood look-alike versions that add warmth, interest, and character to a room. Installation is easier than ever thanks to new, snap-in-place systems—once you have the grid-like tracks installed, you simply pop the panels into place, as illustrated in the drawing below right and described in the sidebar on the opposite page. Also, because these panels are crafted of long-lasting laminate, they don't require any special care. (This is basically the same material used for laminate flooring, but thinner; and here, it doesn't get walked on.) These panels are a great choice for adding flair to a room, whether they're colored a warm wood tone, a white-washed effect, or something in between. Once installed, they may make the ceiling the first thing that people notice.

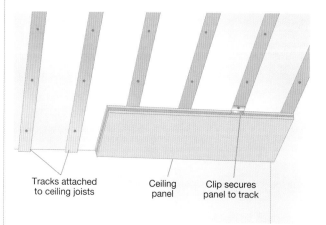

Tracks attached to ceiling joists

Ceiling panel

Clip secures panel to track

LOCATE THE CEILING JOISTS

Ceiling paneling is attached to either furring strips or metal tracks (as shown here). Since both the tracks and the furring strips need to be installed perpendicular to your ceiling joists, your first job is to locate and mark all the joists. Use a stud finder to locate them on opposite walls, as shown. ▼

SNAP CHALK LINES

With the aid of a helper, stretch the line from mark to mark and then snap a chalk line to define each joist location, as shown. ▼

PRO-TIP: SNAP-IN PANELING SYSTEM

Ceiling paneling can be installed several ways. The classic method: Attach furring strips perpendicular to your ceiling joints and then attach the paneling to this via metal clips that are screwed in place, as shown in the drawing below. A much easier way is to use a snap-in track system like the one developed by Armstrong. Their "Easy Up" system uses metal clips as well; but instead of screwing to furring strips, the panels simply snap onto metal tracks that replace the furring strips. ▼

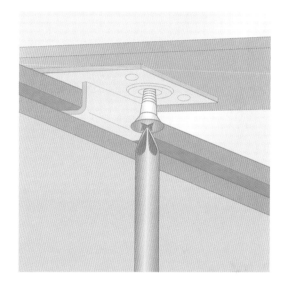

INSTALL THE STARTER TRACK

With the ceiling joists located, you can begin mounting tracks or furring strips. Start in one corner and fasten the track to the joist with nails or screws so the track is about 1" from the wall. You'll find that screws are easier to work with from below and that it's easier to prevent damage to the metal track by screwing it in place versus hammering it in place. Butt the next track up against the first and install it, taking care to maintain 1" distance from the wall, as shown in the bottom left photo.

INSTALL THE REMAINING TRACKS

Once the starter track is in place, measure out 12" and install the next track, as shown in the bottom middle photo. Be careful to offset the joints where two tracks butt up against each other. Since you can secure the tracks only to the ceiling joists, you'll find that some tracks aren't quite as firmly mounted as you'd like. This isn't a problem as long as you offset the joints where the tracks meet. Continue measuring every 12" and installing tracks until you reach the other end of the room. Then install a final track 1" away from the wall.

INSTALL THE STARTER PLANKS

To find out the width of the first row of planks, measure the room parallel to the tracks in inches and divide this by 5 (planks are typically 5" wide). In most cases, your room dimension will not divide evenly. Take the remainder, add 5", and divide this in half; this is the width of your starter planks—and the final planks you'll install on the opposite end of the room from where you started. This will give you a balanced ceiling appearance. Cut enough planks to width, and place the first plank in position. Although you can use clips to support the edge away from the wall, the edge nearest the wall must be screwed to the track. Drill countersunk pilot holes in the plank and attach it to the track with screws, as shown in the bottom right photo.

INSTALL THE STARTER TRACK

INSTALL THE REMAINING TRACKS

INSTALL THE STARTER PLANKS

160

STAGGER THE PLANKS

As you reach the end of the first row, you'll likely need to cut the plank to length. When you start the second row, you'll want to stagger the planks so that the end joints don't line up. Most manufacturers recommend staggering planks in thirds. For example, use a full-length plank for the first row, a two-thirds-length plank for the second row, and a one-third-length plank for the third row, as shown in the top photo. ▶

SNAP THE PLANKS IN PLACE

The beauty of the snap-in paneling system that we used here becomes evident quickly. Just position the next plank, snap a clip on the metal track, and then slide it over until it engages the profiled edge of the plank, as shown in the middle and bottom photos. The ends of the planks are tongue-and-grooved as well, to mate easily. ▶

WORK AROUND ANY FIXTURES

When you encounter a ceiling-mounted fixture, you'll want to do any prep work before adding the paneling. Consult the manufacturer's directions for the recommended offset you'll need to move the electrical boxes. In many cases, this can be accomplished by simply adding a box extender. You can buy box extenders wherever electrical supplies are sold. They can be made of plastic or metal and are designed to extend the outer edge of the box by a set increment, usually ranging from 1/4" to 3/4". The simplest way to work around a fixture is to hold the plank to be installed next to the fixture and mark the fixture location directly onto it. Then cut this out with a saber saw and snap the plank in place. ▼

ADD THE FINAL ROW

Continue installing planks until you reach the opposite wall. If all went well and you made your initial calculations correctly, the final plank should be the same width as your starter plank. Measure the gap between the last plank and the wall and subtract 1/4" for expansion; cut the final plank to this width. Take care to measure and cut each plank individually, since most rooms are not square and you may need

to taper some of these planks slightly. Once you've cut a plank to width, check the fit, as shown. ▼

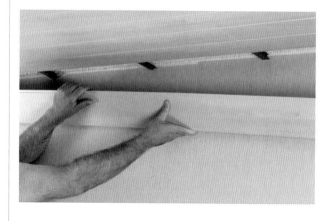

SCREW THE FINAL PLANKS IN PLACE

If your final planks fit, start installing them by slipping the plank's edge into the second-to-last plank. As with the starter row, you'll need to fasten this last row directly to the track that you installed close to the wall. Here again, drill countersunk pilot holes first and then screw each panel in place. Make sure to drill pilot holes as close to the wall as possible so that the molding you'll put up next will conceal the screws. ▼

INSTALL THE PERIMETER TRIM

With all the paneling in place, you can install the perimeter trim to conceal the gap between the paneling and the wall—and to hide the screws used to install the starter and final rows of planks. Some ceiling paneling makers sell matching molding; others do not. You can pick molding that matches (as shown in the bottom photo), or molding that's different, to create an accent. A common trend is to use paint-grade molding and paint it that same color as the accent color used throughout the room. ▼

PRO-TIP: **TIN-LOOK CEILINGS**

Want to add a distinctive touch to a foyer, dining room, or living room? Consider a tin ceiling. If you think it's too much work and too expensive, the market has good news. Take a look at the gorgeous tin ceiling shown here—the surprise is that it's not tin. Instead the tiles are tin-look-alike tiles manu-

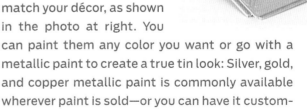

factured by Armstrong as part of their Decorator Ceilings collection. Just like an acoustical tile, this tile goes up quick and easy.

What's really nice is that the tiles all come in white so you can paint them to match your décor, as shown in the photo at right. You can paint them any color you want or go with a metallic paint to create a true tin look: Silver, gold, and copper metallic paint is commonly available wherever paint is sold—or you can have it custom-mixed to suit your needs.

CHAPTER 7:

ARCHITECTURAL TRIM

In days past, architects designed not only homes, but also many of the interior details like trim work and built-ins. This loosely formed what was known as architectural trim. Nowadays, adding these details is often left to the homeowner as contractors try to hold down construction costs. In this chapter we'll show you how to add interesting, elegant, and practical architectural trim, including columns, exterior window and door façades, porch trim, a wall niche, and a fireplace surround.

INSTALLING COLUMNS

Columns can add style to any home's interior or exterior. With the advent of easy-to-use urethane foam versions, columns can readily be incorporated into a room's design. The column we installed here is manufactured by Fypon (www.fypon.com). It's important to note that some decorative foam trim does not provide any structural support; other columns have an internal metal post that offers some support. Check with the manufacturer for stress and load-bearing details.

Columns (sometimes referred to as pillars) can be decorative or structural, round or square, tapered or straight, as illustrated in the drawing below. Decorative columns are made of wood or foam. Structural columns are typically made of solid wood or metal. Plain metal column supports can be wrapped with a decorative wood column, as described on page 169.

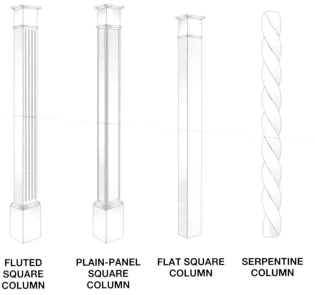

FLUTED SQUARE COLUMN

PLAIN-PANEL SQUARE COLUMN

FLAT SQUARE COLUMN

SERPENTINE COLUMN

LOCATE THE TOP OF THE COLUMN

To install a column or set of columns, start by locating and marking the tops of the columns. How the column installs will depend on the manufacturer. The Fypon column shown here is held in place with a pair of metal retaining plates: one at the top and one at the bottom. ▼

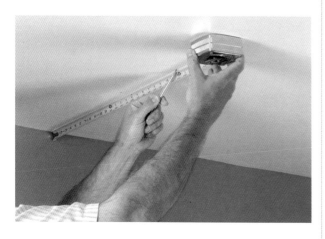

INSTALL THE TOP RETAINER

Center the top retainer plate on the marks you made in the previous step, and mark the mounting hole locations on the ceiling. For any holes that don't lie under a ceiling joist, install plastic anchors for the mounting screws. Then secure the retaining plate to the ceiling. ▼

LOCATE THE BASE RETAINER

With the top retainer plate in place, you can locate the bottom retainer plate. To do this, position the line of a plumb bob in the exact center of the retainer, and then have a helper mark the corresponding point on the floor. ▼

INSTALL THE BASE RETAINER

Position the bottom retainer centered on the mark you just made in the floor, and mark the mounting hole locations on the floor. Then drill pilot holes and attach the bottom retainer. ▼

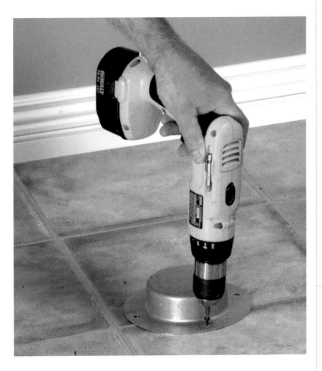

INSTALL THE COLUMN

INSTALL THE COLUMN

With both retainers in place, lift the column and press the top of the column into the top retainer. Swing the bottom over and drop it onto the bottom retainer, as shown in the top right photo.

ADD TOP AND/OR BOTTOM TRIM

All that's left is to add trim to either the top or bottom of the column, or both. In most cases this trim does three things. First, it hides any gaps between the column and the ceiling or floor. Second, if the column mounts on round retainers (as here), the trim prevents the column from rotating. And finally, it creates a smoother transition from the column to the ceiling and/or floor. ▼

PRO-TIP: WRAPPING A METAL COLUMN

Floors and ceilings are often supported by unattractive metal posts or columns—especially in basements. But they don't have to stay that way. It's fairly simple to wrap an existing pole or beam with a wood column, as illustrated in the drawing below. The wrap consists of a plywood sleeve that attaches to the metal pole with a pair of mounting blocks at the top and bottom of the pole. Narrow strips of wood (stiles) are attached to the corners of the box to conceal the plywood edges. To complete the look, top and bottom rails and some quarter-round trim is added. ▼

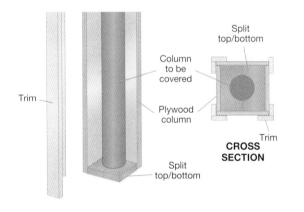

Trim

Column to be covered

Plywood column

Split top/bottom

Split top/bottom

Split top/bottom

Trim

CROSS SECTION

Install the mounting blocks. To wrap a column, start by making a pair of mounting blocks. Size these to fit inside the sleeve that you'll build, and cut a hole in each to match the diameter of the metal pole. Cut each block in half and position the split block around the pole. Then secure the blocks to the floor and to the ceiling, as shown. ▶

Install the wrap and trim. Cut plywood sleeve parts to width and length so they're slightly shorter than your floor-to-ceiling measurement. Then fasten together two sides and a back to create a U-shaped sleeve. Slip the sleeve over the mounting blocks and fasten it to the blocks. Now add the cover to the sleeve, as shown in the bottom inset photo. Attach the stiles to the corners of the sleeve with glue and nails, as shown in the photo below. Finally, miter some quarter-round or cove molding to wrap around the top and bottom of the column to conceal any gaps between the column and the floor or ceiling. ▼

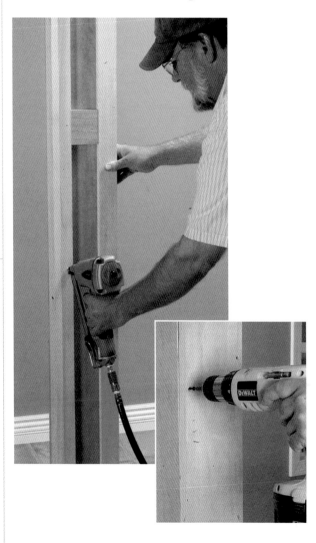

EXTERIOR WINDOW FAÇADE

Tired of your home's plain exterior? Dress it up with new window façades. What's really nice about the exterior trim we used here is that it's easy to install and virtually impervious to weather. How is this possible? The trim is urethane foam,

manufactured by Fypon (www.fypon.com). It cuts easily and goes up with a bead of polyurethane adhesive and a couple of fasteners. Foam trim is available in a huge variety of shapes, sizes, and styles, from simple molding to arched or decorative window panels, as illustrated below. ▼

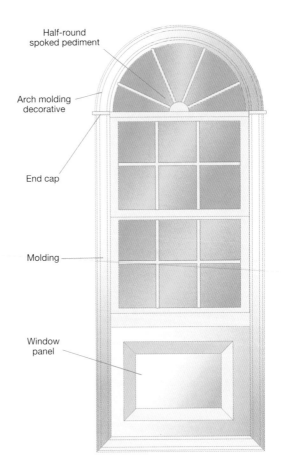

Half-round spoked pediment

Arch molding decorative

End cap

Molding

Window panel

APPLY AN ADHESIVE

Foam trim should always be secured with both an adhesive and fasteners. For exterior-mounted trim, apply a generous bead of a quality polyurethane adhesive. Note: If you're mounting the trim to a home covered with vinyl siding, read and follow the manufacturer's mounting instructions. Quite often they'll have you attach the trim without glue, then remove the trim and enlarge the screw holes in the siding before reattaching the trim with adhesive. Enlarging the holes like this prevents problems with seasonal expansion and contraction of the siding. ▼

ATTACH THE SIDE WITH GALVANIZED NAILS

To install exterior window trim, first remove any existing trim (if applicable). Then follow the manufacturer's directions to measure and cut the side trim to size. Apply a bead of polyurethane adhesive as described above and secure each side with fasteners, as shown in the top right photo.

ATTACH THE SIDE WITH GALVANIZED NAILS

ATTACH THE HEADER

With the sides in place, cut the header to size (if necessary) and attach it with polyurethane adhesive and fasteners. Note: If you do have to cut the header to fit, cut it so the seam ends up in the middle of the header. This way it can be concealed with a keystone, as described on page 172. ▼

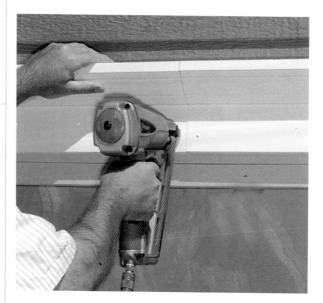

ADD THE KEYSTONE

Although keystones were originally used to lock together the stones of a masonry arch, a foam keystone is purely decorative. In addition to adding a nice touch to a window façade, the keystone can also conceal a seam if the header had to be cut to fit. As with any other foam trim, attach the keystone with polyurethane adhesive and exterior-rated fasteners. ▼

ADD THE APRON

INSTALL THE ROSETTES

The final step is to install the rosettes with polyurethane adhesive and galvanized fasteners. Depending on where these are positioned over your exterior covering, you may or may not need to insert shims behind the rosettes to make them flush with the side trim and apron. Fill all nail holes with an exterior-rated putty, sand smooth, and paint as desired. ▼

ADD THE APRON

Next, cut the apron to length and secure with polyurethane adhesive and fasteners, as shown in the top right photo. If you're using rosettes (as we did here), you'll cut the apron to fit between the inside bottom corners of the side trim. If you're not using rosettes, the apron should come with finished ends and you'll need to cut it as you did the header, making sure the seam ends up in the middle of the apron.

EXTERIOR DOOR FAÇADE

Almost every exterior door can look better with trim added. Just look at the amazing difference between the before and after photos shown below. Although it looks like the wood trim was hand-cut and installed by a master craftsman, it's made of foam. And foam trim like this is super-easy to install and virtually impervious to weather. The trim shown here is made by Fypon (www.fypon.com). It cuts easily and goes up with a bead of polyurethane adhesive and a couple of fasteners. Foam trim is available in a variety of shapes, sizes, and styles, from simple molding to arched or decorative side panels, as illustrated in the drawing at right.

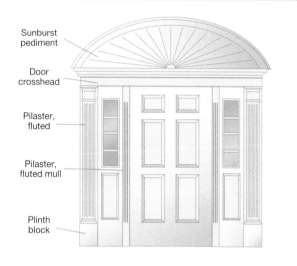

Sunburst pediment

Door crosshead

Pilaster, fluted

Pilaster, fluted mull

Plinth block

INSTALL THE SIDE TRIM

The foam trim we installed here is designed to butt up against existing brick mold. To install it, start by measuring and cutting the side pilasters to length. The ones shown here run from the top of the brick mold to its bottom. Apply a bead of polyurethane adhesive to the back of the pilaster, align it with the top edge of the brick mold, and fasten it in place with galvanized nails. ▼

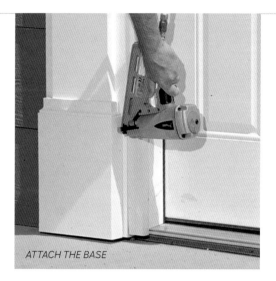

ATTACH THE BASE

INSTALL THE HEADER

The next step is to install the header. The foam header we installed comes long and has to be cut to length—basically you have to remove a center section, leaving the desired-length header with its seam in the center. To prevent water from becoming trapped behind the header, the manufacturer suggests that you install metal flashing (called drip edge) under the siding above the header. The flashing fits over the header and creates a watertight seal (see the manufacturer's installation instructions for more on this). Once you've got the header cut to size, install the first half. ▼

ATTACH THE BASE

The side pilasters we used are two-piece units consisting of a pilaster and a base. The base fits over the cut end (bottom) of the pilaster and provides a finished look. Install a base flush with the bottom of each pilaster by applying a bead of polyurethane adhesive and driving in fasteners from the side, as shown in the top right photo.

INSTALL SECOND HALF OF THE HEADER

Apply a bead of polyurethane adhesive to the end of the remaining header portion, and butt it up against the installed half. Then secure this to the exterior wall covering with adhesive and galvanized fasteners, as shown. ▼

ADD THE KEYSTONE

To conceal the seam in the header, install the keystone with adhesive and exterior-rated fasteners, as shown in the top right photo. In masonry, a keystone is the central wedge-shaped stone of an arch that locks its parts together. In house trim, a keystone can be purely decorative and is installed on both arched and straight headers. In addition to being decorative, a keystone can be used to conceal an underlying joint, like the joint between the two foam header pieces shown here. Fill all nail holes, and apply the exterior finish of your choice.

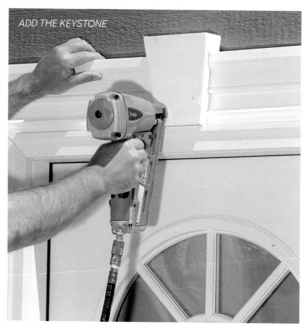

ADD THE KEYSTONE

PRO-TIP: MEDALLIONS

If you can't find the trim of your dreams for your exterior door, consider making your own. It's easy to make your own decorative pilasters. Use strips of wood for the sides and top and make your own rosettes. To do this, cut square blocks slightly larger than the width of your side/top trim. Then attach pre-made medallions to the blocks. Medallions (like those shown below) can be found wherever trim is sold and also at craft stores. Just make sure to attach them to the blocks with exterior-rated adhesive and fasteners. ▼

175

EXTERIOR PORCH TRIM

The porches on some older homes and many new homes consist of nothing more elaborate than a roof supported by a couple of columns and either a wood or a concrete floor. What about a railing? On a porch, it's like shutters on a window: not absolutely essential, but it makes the home look more finished. A railing also provides additional privacy, and can even improve safety by offering a handhold, and by creating a boundary for small children and pets.

A railing can be as elaborate as sculpted posts and balusters, as illustrated in the top right drawing, or as simple as the wood railing shown in the bottom right photo. Our railing consists of a simple 2x4 frame, a set of 2x2 balusters, and a 1x6 cap rail. Additionally, you can add visual interest (and maybe a touch of whimsy) to a porch with ornamentation. This can be anything from simple trim molding to fancy decorative scrollwork, as shown in the top left photo. Finally, if the columns supporting your roof are metal or are just plain boring, consider dressing them up with a wood wrap, as described on page 169.

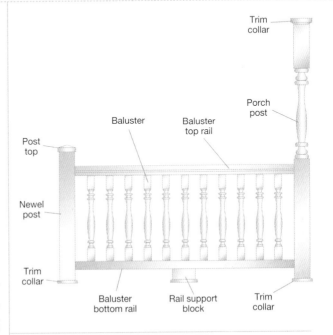

LAY OUT THE BALUSTERS

To make a wood railing to fit between a pair of columns, start by measuring the distance between the columns and cut a pair of top/bottom rails to this measurement. Then place the rails side by side with the ends flush, and lay out the desired baluster spacing. This number should not exceed 5" (check your local building code for more on this). Mark a centerline on both rails and lay out the balusters, working from the center out toward the ends, as shown. ▼

ASSEMBLE THE FRAME

Before you can assemble the frame, you'll need to cut the end pieces to fit between the rails. How long these are will depend on the length of your balusters and how far you want the balusters to protrude past the bottom rail. Once you've decided on the length, cut the ends to length. Then assemble the frame by driving in pairs of galvanized deck screws through the rails and into the ends, as shown in the top right photo. Since you'll be driving these screws in near the ends of the rails, it's best to first drill pilot holes.

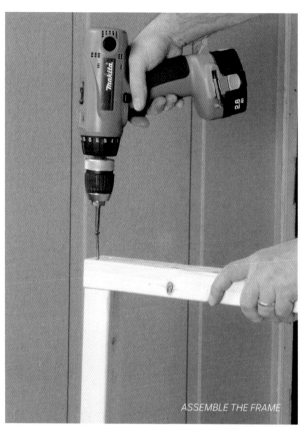

ASSEMBLE THE FRAME

INSTALL THE BALUSTERS

Now you can attach the balusters. Align each baluster with the lines that you laid out earlier and attach them with 2½" galvanized screws, as shown. Again, it's best to drill pilot holes first. ▼

ATTACH THE RAILING

With the balusters in place, you can attach the assembled railing to the porch columns. Gently slide the railing in place at the desired distance up from the porch floor. Then use a level to make sure the top is level before screwing through the ends of the frame and into the columns. Make sure to use 2½" or 3" galvanized deck screws for this, spacing them out every 12" or so. ▼

ATTACH THE RAIL CAP

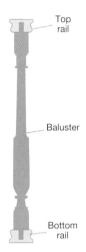

ATTACH THE RAIL CAP

Although you can leave the railing as it is, the frame assembly screws will be exposed. Not only would this be unattractive, but it would also cause problems over time, as moisture wicks along the screw threads and encourages rust to develop. A nice finishing touch is to add a 1x6 cap rail. Measure the distance from column to column, and cut a cap to fit. If you have access to a router, consider easing the top edges by routing a ¼" or ⅜" round-over. Secure the cap rail to the frame assembly by screwing up through the frame with 2" galvanized screws. If your drill won't fit between the balusters, secure the cap rail from above but drive the screw below the surface of the cap (as shown in the top right photo). Then fill these holes with exterior-grade putty.

PRO-TIP:
INSTALLING BALUSTERS

If you'd like to add style to your railing, consider purchasing pre-made balusters. They come in many sizes and shapes. as illustrated in the drawing below. Also, how you mount the balusters to the top and bottom rail will affect the overall appearance. The most common way to do this is to capture the baluster between the top and bottom rails, as illustrated in the drawing at right. Alternatively, they can be attached to the sides of the rails, as we did here. ▼

Top rail

Baluster

Bottom rail

ADD A SIMPLE PLANTER

A classic way to add a splash of color to a porch is with flowers. Although you can place plants in a container on the floor, they'll be a lot more visible in a planter that rests on a rail. Our railing planter consists of a front and back, two sides, and a bottom—all cut from 1x 6 pine. (Note: Cedar and redwood are both excellent choices for this project, since they are naturally weather- and rot-resistant.) The sides are $5\frac{1}{2}$" wide, and the front and back can be cut to any length. The bottom is also a length of 1x6, cut to fit inside the box once it's assembled. Cut the parts to size, and assemble with exterior-rated glue and fasteners. Place the planter on the railing in the desired location so the planter is centered, and secure it to the railing by driving 2" galvanized deck screws through the bottom of the planter and into the railing, as shown. ▼

DRILL DRAINAGE HOLES

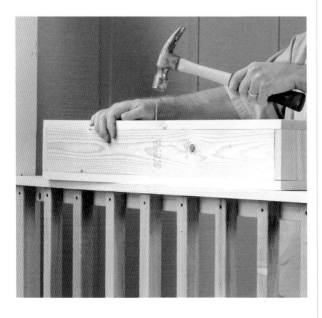

DRILL DRAINAGE HOLES

To prevent the planter from filling with water from rain or overwatering, drill a series of drainage holes in the bottom, as shown in the top right photo. An easy way to waterproof the inside of the planter is to coat the inside with roofing cement.

PAINT THE RAILING

To protect your new porch railing, you should apply some type of finish. If you used dissimilar woods (as we did here), it's best to paint the railing for a uniform color, as shown. Alternatively, you can apply an exterior stain, clear finish, or waterproofing finish. ▼

Scrollwork is an excellent way to spice up an otherwise drab exterior. You can purchase pre-made scrollwork, like the urethane ornamentation shown here, manufactured by Fypon (www.fypon.com). Or you can make your own, as described on page 181.

APPLY ADHESIVE

Foam scrollwork should never be installed with just fasteners. It's designed to be installed with both fasteners and a bead of high-quality, polyurethane adhesive. You don't need a lot here: About a $1/8$" bead on both mating surfaces will work fine. You'll find that this combination of adhesive and fasteners along with the foam's built-in flexibility will make installing scrollwork a snap. ▼

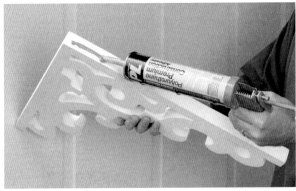

MARK THE LOCATION

Step one for installing ornamentation is to locate and mark its location. A helper is useful for this since it's hard to determine the visual impact of molding when you're right next to it. Place the scrollwork where you think it should go, and have someone look at the house from the sidewalk or street. Adjust as needed, and when you've located the perfect spot, use a pencil to mark its location on the house, as shown. ▶

SECURE THE ORNAMENTATION

To install ornamentation, press it in place, taking care to align its edges with the pencil marks that you made earlier, and attach it with exterior-rated fasteners. Then go back and fill in nail holes with an exterior-rated putty and paint it if desired. ▶

PRO-TIP:
MAKING SCROLLED ORNAMENTATION

It's easy to make your own scrollwork if you have a scroll saw or saber saw. You can create your own pattern or use the one shown here. For long-lasting scrollwork we recommend that you use a weather-resistant wood for this, like cedar or redwood.

Transfer the pattern. The first step to making a set of scrolled ornamentation is to transfer the pattern to the wood blanks. For intricate patterns like the one shown here, it's often easiest to make paper copies of the pattern and attach the pattern directly to the blank with some type of adhesive. Usually, a spray-on adhesive is applied to the back of the pattern and then it's pressed in place on the workpiece, as shown in the top right photo. Alternatively, if the pattern is simple, you can cut it out, hold it in place on the workpiece, and trace around it with a pencil.

Cut out the shape. For the pattern shown here, you'll need to drill access holes for your scroll or saber saw blade to make the "pierced cuts." Drill as many as you like, to make cutting the pattern as easy as possible. To make a pierced cut, insert the blade in an access hole and guide the workpiece with both hands, as shown in the bottom photo. When following a pattern, you should either try to split the pattern line with the saw blade or stay just a hair to the waste side. Staying to the waste side of the line lets you come back with a sanding stick or file later and sand or file exactly to the line. Round over any sharp edges, and paint as desired.

TRANSFER THE PATTERN

CUT OUT THE SHAPE

INSTALLING A NICHE

Want an "instant" makeover for a wall? Consider installing a pre-made wall niche like the one shown here. Not only does a wall niche offer distinctive display space, but it can also serve as a handy shelf in an entryway for keys, mail, etc. Although you'd never know it by looking, the wall niche shown here is made of urethane foam, manufactured by Fypon (www.fypon.com). This particular niche is designed to fit between studs, so installation is a snap. ▼

LOCATE THE WALL OPENING

Use a stud finder to locate the wall studs where you'll be locating the wall niche. Use a pencil to mark each stud location. Note that if you want to shift the niche to one side or the other of the studs, you need to take precautions to avoid weakening the wall. You'll have to remove the drywall, cut the studs, and install a header and sill plate. The next step is to mark the opening in the wall for the niche. Some manufacturers provide a paper template for this; others don't. If your niche didn't come with a pattern, flip the niche over and measure its back; then transfer these measurements to the wall, as shown. ▼

CUT OUT THE OPENING

Once you've marked the opening in the wall, cut out the opening. You can do this with a drywall saw, as shown here, or with a reciprocating saw. ▼

INSTALL THE NICHE

With the opening cut, test the fit of the niche, as shown in the top right photo. If you cut the opening to match a template provided by the manufacturer, it will probably fit. If you made your own template or transferred measurements, it may not fit; simply tweak the opening a bit. On the niche shown here, there's a large rim around its perimeter. This is an excellent place to apply a bead of polyurethane glue before inserting the niche into the opening; make sure the rim presses flat against the wall.

SECURE THE NICHE

Although the polyurethane glue is probably all you'll need to secure the niche to the wall, it's a good idea to drive a couple of nails through the inside of the niche into the wall studs, as shown in the bottom photo. Fill these holes and apply the finish of your choice.

INSTALL THE NICHE

SECURE THE NICHE

INSTALLING A FIREPLACE SURROUND

A common trend used in new construction to keep costs down is to minimize the trim work, which is both labor-intensive and expensive. One particular area that's often left plain is a fireplace. Quite often there is no trim at all, as with the one shown in the bottom photo. Look what a difference the addition of a fireplace surround makes (top photo).

You can build a fireplace surround from scratch, but that takes considerable woodworking skills—and equipment. A more homeowner-friendly approach is to use a fireplace surround kit, like the one illustrated below and described here. The kit we used is made by Decorative Concepts (www.decorativeconcepts.net). The kit consists of a mantel, two columns, and a pair of column wraps; see the opposite page for more on surround kit options.

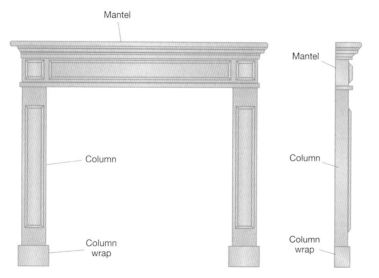

Mantel

Column

Column wrap

Mantel

Column

Column wrap

PRO-TIP: SURROUND KITS

Surround kits come in two basic flavors, homeowner's and builder's, as illustrated in the drawing at right. The most user-friendly option is the homeowner's kit. It consists of a mantel, columns, and column wraps. The easy-to-assemble parts are bolted securely together and attach to the wall via a cleat. The cleat is secured to wall studs, and the surround is screwed to the cleat.

In the less expensive builder's kit, more work is left for a homeowner. The columns and mantel are pre-made, but the molding is left off. This means that you'll need to miter-cut and install crown molding and add additional accent strips.

Both types of kits are available in MDF (medium-density fiberboard) and can be painted as desired. Alternatively, you can choose from a variety of solid-wood mantels and can either stain them or apply a clear finish. Check with your surround manufacturer for specifics on measuring your fireplace so that you can order a correctly sized surround kit.

HOMEOWNER'S KIT

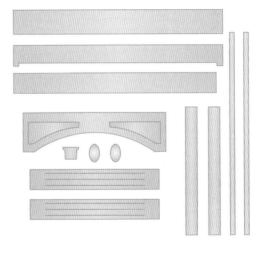

BUILDER'S KIT

185

MEASURE THE HEIGHT

Depending on the size of your fireplace, you may or may not be able to order a surround that will fit without alterations. On others, you may need to measure the height of your fireplace and cut the columns to size. See the manufacturer's installation directions for specifics on calculating the correct height of the columns. ▼

ASSEMBLE THE MANTEL

Because the mantel itself can be quite heavy, the easiest way to assemble the surround is to start with the mantel upside down on the floor, as shown in the top right photo. Position columns at both ends of the mantel, and secure them with the bolts provided.

ASSEMBLE THE MANTEL

LOCATE THE SUPPORT CLEAT

With the aid of a helper, flip the surround upright, taking care to support the columns as you do this. Position the surround so it's centered on the fireplace, as shown. Then use a pencil to lightly scribe a line on the wall where the top back edge of the mantel rests against it. Remove the mantel and measure down from this line the thickness of the mantel top. Then draw a level line to locate where the support cleat will be installed. ▼

LOCATE THE WALL STUDS

Now use an electronic stud finder to locate the wall studs along the line you marked in the previous step. Mark each stud location. This is where you'll attach the support cleat. ▼

ATTACH THE SUPPORT CLEAT

Hold the support cleat in place so its top edge is flush with the line you drew earlier. Then drill pilot holes through the cleat and into the wall stud. Secure the support cleat to the wall studs with the long screws provided. ▼

SECURE THE MANTEL TO THE SUPPORT CLEAT

Now you can position the surround centered on the fireplace, with the mantel top resting on the support cleat. Use the trim-head screws provided to secure the mantel to the support cleat. ▼

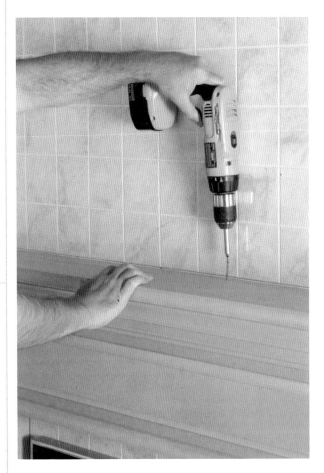

SECURE THE COLUMNS TO THE WALL

With the mantel secured to the wall, the next step is to secure the bottoms of the columns to the wall. Use the long screws provided and drive them through the column wraps and into the wall—they should penetrate into the bottom plate of the wall. Note that if you had to trim the columns to length, you'll now need to drill counterbored holes near the bottom of each column to replace the mounting holes that you trimmed off. ▼

ADD THE COLUMN WRAPS

All that's left is to slide the column wraps over the base of the columns and secure them with a few nails, as shown in the top right photo. Fill all nail and screw holes, and apply the finish desired.

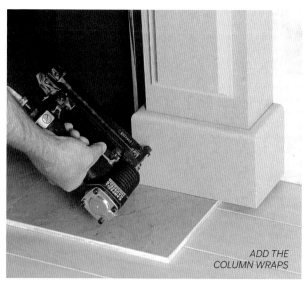

*ADD THE
COLUMN WRAPS*

PRO-TIP: ONE-PIECE SURROUNDS

An alternative to using a surround kit is to purchase a pre-made one-piece surround. The one-piece surround shown here is made of urethane foam and is sold by Fypon (www.fypon.com). The disadvantage to a single-piece surround is that it has to fit your fireplace—unlike most wood kits, it can't be easily modified. ▼

PRO-TIP: **MAKING ROSETTES**

If you're adding a fireplace surround or just want to dress up an existing surround, consider adding rosettes. Rosettes are decorative wood blocks installed at the upper and/or lower corners of trim. They can be mounted to a block, be carved into a block, or come as a separate appliqué that you can then attach to a block or existing trim (see below). ▼

Appliqués. Rosette appliqués come in a huge variety of sizes and patterns, as illustrated in the drawing below and in the bottom left photo. They can be found wherever trim is sold, and at most craft stores.

Attach appliqué to a blank. If you want to make your own rosettes, start by cutting square blocks of wood to the desired size, or buy pre-made blocks. Apply glue to the back of the appliqué, and center it on the block. Then drive in a couple of fasteners, as shown in the bottom photo. Your rosettes are now ready to use. ▼

APPLIQUÉS

INDEX

A

Adhesives, 24
Air tools
 described, 40–42
 fasteners for, 44
 safety rules for, 77
 troubleshooting,
  78–79
 using, 76–77
Angled stops, 68
Angle gauges, 29, 59
Angle squares, 29, 54
Awls, 48

B

Backer strips and panels,
 104
Back saws, 32
Balusters, pre-made, 178
Base, for crown molding,
 73, 150
Baseboard
 described, 11, 14
 installing, 92–97
Bevels, cutting, 87
Biscuit jointers, 37
Brackets
 pre-made, 120
 shop-made, 122–123
Brad nailers, 41. *See also*
  Air tools
Brads, 43, 74
Built-up trim
 advantages of, 11, 17
 installing, 96
 making, 88
Burnishing, 154

C

Cabinet trim
 crown molding, 118–119
 door frames, 116–117
Casing. *See* Door trim;
  Window trim
Cat's paws, 89
Caulk, 23, 85
Caulk saver rod, 23
Ceiling joists, finding, 72
Ceiling trim. *See also*
  Crown molding
 foam, 147, 157
 install sequence for,
  144
 medallions, 10, 148
 paneling, 158–163
 period trim, 10
 simple molding,
  145–146
 tin-look, 163
Chair rail
 described, 15
 installing, 98–99
Chalk lines, 31
Chip-out, preventing,
 69
Chisels, 34
Circular saws, 35, 62
Columns
 installing, 167–168
 types of, 166
 wrapping, 169
Combination squares
 described, 30
 using, 50–51, 131
Compasses, 30

Cope-and-stick joints,
 111
Coped joints, 70–71
Coping saws, 32, 71
Corners
 baseboard, 97
 crown molding, 154
 pre-made, 22, 95, 157
 wainscoting, 106
Cove cutting, 87
Creep, preventing, 68, 69
Crown molding
 on cabinets, 118–119
 cutting, 38, 67, 152–153,
  155
 described, 10, 15
 design guidelines for,
  149
 foundations for, 73, 150
 installing, 151–154
 two-piece, 156
Custom trim, 86–88
Cutting techniques
 circular saw, 62
 crown molding, 38, 67,
  152–153, 155
 miter box, 61

D

Decorative trim, 8, 16
Demolition, 89
Digital protractors, 29, 60
Door stops, 16
Door trim
 described, 12, 15
 on door openings,
  140–141

exterior façades,
  173–175
 installing, 136–139
 nailing, 130
 reveals, 131
Drills and drilling, 39, 72
Duplicate parts, 122

E

EasyCoper, 35
Expanding foam, 23
Exterior trim
 caulking, 85
 doors, 138–139,
  173–175
 porch, 176–180
 windows, 134–135,
  170–172

F

Fasteners, 43–44
Files, 34, 80
Finger gauge, 131
Finishing supplies, 25
Fireplace surrounds
 described, 13
 installing, 186–188
 kits, 184–185
 one-piece, 188
Floor trim. *See* Base-
  board
Flutes, routing, 127
Foam trim
 ceiling trim, 147, 157
 described, 19
 exterior façades,
  170–175

190

fireplace surrounds, 188
pilasters, 124
scrollwork, 180
wall niches, 182–183
Folding rules, 28, 49
Framing members, finding, 72
Framing squares, 30, 55
Function, of trim, 8–9. *See also specific types*

G
Gaps and openings, 9
Glues, 24
Golden rectangle, 100

H
Hammers, 39, 74–75
Hardboard, 21
Hardwood, 20

I
Interchangeable trim, 17
Interior trim, caulking, 85. *See also specific types*

J
Jigs, crown molding, 38, 67, 153
Joinery
  cope-and-stick joints, 111
  coped joints, 70–71
  scarf joints, 94

K
Kerf inserts, 69
Keystones, 172, 175

L
Laminated trim, 19
Laser guides, 38, 65

Laser levels, 31, 98
Layout and measuring. *See also specific tools*
  techniques for, 58
  tools for, 28–31
Levels, 28, 31, 98
Lock-nailing, 75, 137
Lumber types, 20

M
MDF, 18, 21
Medallions
  ceiling, 10, 148
  for custom trim, 175
Medium-density fiber-board (MDF), 18, 21
Miter boxes
  described, 32
  using, 61, 81
Mitered returns, 94, 132
Miters
  lock-nailing, 75, 137
  trimming, 80–81
Miter saws. *See* Power miter saws

N
Nail guns. *See* Air tools
Nail holes, 25, 82–84
Nailing blocks, 73, 150
Nailing techniques
  basic, 74–75
  for window and door trim, 130, 137
Nails
  described, 43–44
  removing, 89
Nail sets, 39, 75
Needle-nose pliers, 74
Niches, 182–183

P
Paint, 25
Paint-grade trim, 18

Pattern-routing, 122–123
Period trim, 10, 11
Picture frame casing
  described, 12
  installing, 132–133
Picture rail, 120–123
Pilasters, 124–127
Planes, 33, 80
Planters, 179
Plastic trim, 19
Plate rail, 120–123
Playing card shims, 68, 81
Plinth blocks, 22, 140
Plywood, 21
Porch trim
  railing planters, 179
  railings, 176–178
  scrollwork, 180–181
Power miter saws
  accessories for, 38, 65, 67
  basic use of, 63–64
  coping with, 70
  described, 37
  laser guides for, 65
  safety rules for, 64
  tips for using, 68–69, 81
  workpiece positions for, 66
Power tools
  described, 35–38
  using, 62–71
Practical trim, 9, 16
Putty, 25, 83–84
Putty knives, 39

R
Railings, 176–178
Raised panels, 110–112
Rasps, 34, 80
Reciprocating saws, 37
Removal of trim, 89
Reveals, marking, 131

Riflers, 34
Rosettes
  described, 22
  installing, 141
  making, 189
Router bits, 86
Routers, 36
Routing techniques
  custom trim, 87–88
  flutes, 127
  pattern-routing, 122–123
  raised panels, 110–111
Rule stops, 58

S
Saber saws, 35, 181
Safety rules, 64, 77
Sandpaper, 69
Saws. *See also specific types*
  hand-powered, 32
  power tools, 35–38
Scarf joints, 94
Screen stops, 16
Screws, 45
Scribing, 56–57
Scroll saws, 71, 181
Scrollwork, 180–181
Sealants, 23
Sheet goods, 21
Shims
  for nailing brads, 74
  for saws, 68, 81
Shooting boards, 33, 81
Softwood, 20
Spring angle, 149, 155
Squares
  described, 29, 30
  using, 50–51, 54–55, 131
Square stops, 68
Stain-grade trim, 18
Stains, 25
Staplers, narrow-crown, 41

Stickering, 105
Stool-and-apron casing
  described, 12
  installing, 132–133
Stop blocks, 68
Story sticks, 52–53
Stud finders, 31, 72
Support, for workpiece,
  69

T
Table saws
  described, 36
  using, 87, 112
Tape measures, 28, 48
Tools. *See also specific
    types*
  air-powered, 40–42

  for assembly, 39
  fasteners, 43–44
  hand tools, 32–34
  layout and measuring,
    28–31
  power tools, 35–38
  Trim foundations, 73
  Try squares, 30, 50–51

W
Wainscoting
  conditioning, 105
  described, 13, 14
  frame-and-panel,
    110–115
  installing, 102–107
  quick-install,
    108–109

Wall frames
  described, 13
  installing, 100–101
Wall niches, 182–183
Wall trim, 13. *See also
    specific types*
Wax crayons, 25, 82
Window-and-door
    casing. *See* Door
    trim; Window trim
Window trim
  described, 12, 15, 132
  exterior façades,
    170–172
  installing, 132–135
  nailing, 130
  reveals, 131

192

Metric Equivalency Chart
Inches to millimeters and centimeters

INCHES	MM	CM	INCHES	CM	INCHES	CM
1/8	3	0.3	9	22.9	30	76.2
1/4	6	0.6	10	25.4	31	78.7
3/8	10	1.0	11	27.9	32	81.3
1/2	13	1.3	12	30.5	33	83.8
5/8	16	1.6	13	33.0	34	86.4
3/4	19	1.9	14	35.6	35	88.9
7/8	22	2.2	15	38.1	36	91.4
1	25	2.5	16	40.6	37	94.0
1 1/4	32	3.2	17	43.2	38	96.5
1 1/2	38	3.8	18	45.7	39	99.1
1 3/4	44	4.4	19	48.3	40	101.6
2	51	5.1	20	50.8	41	104.1
2 1/2	64	6.4	21	53.3	42	106.7
3	76	7.6	22	55.9	43	109.2
3 1/2	89	8.9	23	58.4	44	111.8
4	102	10.2	24	61.0	45	114.3
4 1/2	114	11.4	25	63.5	46	116.8
5	127	12.7	26	66.0	47	119.4
6	152	15.2	27	68.6	48	121.9
7	178	17.8	28	71.1	49	124.5
8	203	20.3	29	73.7	50	127.0

mm = millimeters   cm = centimeters